The *gyroid*. An infinitely connected periodic minimal surface containing no straight lines, recently discovered and christened by A. H. Schoen. (NASA Electronics Research Laboratory)

A Survey of
MINIMAL SURFACES

by

ROBERT OSSERMAN

Stanford University

DOVER PUBLICATIONS, INC.
NEW YORK

Published in Canada by General Publishing Company, Ltd., 30 Lesmill Road, Don Mills, Toronto, Ontario.
Published in the United Kingdom by Constable and Company, Ltd., 10 Orange Street, London WC2H 7EG.

This Dover edition, first published in 1986, is a corrected and enlarged republication of the work first published in book form by the Van Nostrand Reinhold Company, New York, 1969. (A fundamentally identical version had been translated into Russian and published in the journal *Uspekhi Matematicheskikh Nauk* in 1967.) A new Preface, an Appendix updating the original edition, and a supplementary bibliography have been specially prepared for this edition by the author, who has also made several corrections.

Manufactured in the United States of America
Dover Publications, Inc., 31 East 2nd Street, Mineola, N.Y. 11501

Library of Congress Cataloging in Publication Data

Osserman, Robert.
  A survey of minimal surfaces.

  Bibliography: p.
  Includes index.
  1. Surfaces, Minimal. I. Title.
QA644.O87 1986   516.3'62   85-12871
ISBN 0-486-64998-9 (pbk.)

PREFACE TO THE DOVER EDITION

The flowering of minimal-surface theory referred to in the Introduction has continued unabated during the subsequent decade and a half, and has borne fruit of dazzling and unexpected variety. In the past ten years some important conjectures in relativity and topology have been settled by surprising uses of minimal surfaces. In addition, many new properties of minimal surfaces themselves have been uncovered. We shall limit ourselves in this edition to a selected updating obtained by enlarging the bibliography (in "Additional References") and by the addition of "Appendix 3," where a few recent results are outlined. (The reader is directed to relevant sections of Appendix 3 by new footnotes added to the original text.) We shall concentrate on results most closely related to the subjects covered in the main text, with the addition of some particularly striking new directions or applications. Fortunately we can refer to a number of survey articles and books that have appeared in the interim, including the encyclopedic work of Nitsche [II] and, for the approach to minimal surfaces via geometric measure theory, the Proceedings of the AMS Symposium—Allard and Almgren [I]; both books include an extensive bibliography. The section "Additional References," following the original bibliography in the present work, starts with a list of those books and survey articles where many other aspects may be explored.

In this Dover edition, a number of typographical errors have been corrected, and incomplete references in the original bibliography have been completed. Otherwise, with minor exceptions, the original text has been left unchanged.

NOTE: In references to the bibliography, Roman numerals refer to the list of books and survey articles in the Additional References (pp. 179–200), while Arabic numerals refer either to the subsequent list of research papers or to papers in the original References (pp. 167–178). *MSG* and *SMS* refer to *Minimal Submanifolds and Geodesics*, the proceedings of a conference held in Tokyo in 1977, and to *Seminar on Minimal Submanifolds*, a collection of papers presented during the academic year 1979–1980 at the Institute for Advanced Study, listed as the first and second items in the Additional References, under books and survey articles.

ROBERT OSSERMAN

PREFACE TO THE FIRST EDITION

This account is an English version of an article (listed as Item 8 in the References) which appeared originally in Russian. In the three years that have passed since the original writing there has been a flurry of activity in this field. Some of the most striking new results have been added to the discussion in Appendix 2 and an attempt has been made to bring the references up to date. A few modifications have been made in the text where it seemed desirable to amplify or clarify the original presentation. Apart from these changes, the present version may be considered an exact "translation" of the Russian original.

CONTENTS

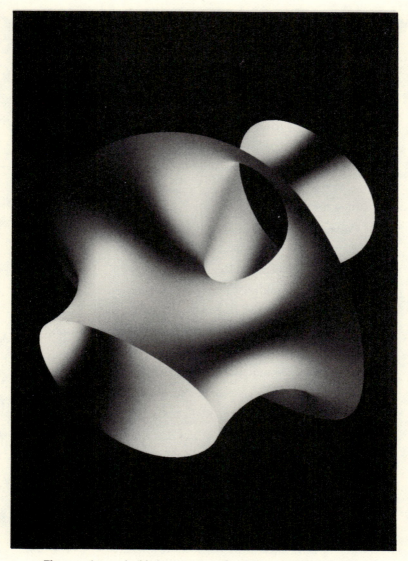

The complete embedded minimal surface of Costa-Hoffman-Meeks
(see Appendix 3, Section 4). (Photograph copyright © 1985 by
D. Hoffman and J. T. Hoffman.)

Meeks: 413-545-4239   U.Mass

D. Hoffman: 413-545-0386 U. MASS
     413-545-2812 Sec
     413-549-6575 (H)

INTRODUCTION

The theory of minimal surfaces experienced a rapid development throughout the whole of the nineteenth century. The major
achievements of this period are presented in detail in the books of
Darboux [1] and Bianchi [1]. During the first half of the present
century, attention was directed almost exclusively to the solution
of the Plateau problem. The bulk of results obtained may be found
in the papers of Douglas [1, 2], and in the books of Radó [3] and
Courant [2]. A major exception to this trend is the work of Bernstein [1, 2, 3] who considered minimal surfaces chiefly from the
point of view of partial differential equations. The last twenty
years has seen an extraordinary flowering of the theory, partly in
the direction of generalizations: to higher dimensions, to Riemannian spaces, to wider classes of surfaces; and partly in the direction of many new results in the classical case.

Our purpose in the present paper is to report on some of the
major developments of the past twenty years. In order to give any
sort of cohesive presentation it is necessary to adopt some basic
point of view. Our aim will be to present as much as possible of
the theory for two-dimensional minimal surfaces in a euclidean
space of arbitrary dimension, and then to restrict to three dimensions only in those cases where corresponding results do not seem
to be available. For a more detailed account of recent results in
the three-dimensional case, we refer to the expository article of

Nitsche [4], where one may also find an extensive bibliography and a list of open questions. The early history of minimal surfaces in higher dimensions is described in Struik [1].

Since it is impossible to achieve anything approaching completeness in a survey of this kind, we have selected a number of results which seem to be both interesting and representative, and whose proofs should provide a good picture of some of the methods which have proved most useful in this theory. For the convenience of the reader, a list of the theorems proved in the paper is given in Appendix 1.

For the most part the present paper will contain only an organized account of known results. There are a few places in which new results are given; in particular, we refer to the treatment of non-parametric surfaces in E^n in Sections $2-5$, and the discussion of the exterior Dirichlet problem for the minimal surface equation in Section 11.

In Appendix 2 we try to give some idea of the various generalizations of this theory which have been obtained in recent years.

One word concerning the presentation. In most treatments of differential geometry one finds either the classical theory of surfaces in three-space, or else the modern theory of differentiable manifolds. Since the principal results of this paper do not require any knowledge of the latter, we have decided to give a careful introduction to the general theory of surfaces in E^n. For similar reasons we have included a section on the simplest case of Plateau's problem in E^n, for a single Jordan curve. In this way we hope to provide a route which may take a reader with no previous knowledge of the theory, directly to some of the problems and results of current research.

§1. *Parametric surfaces: local theory*

We shall denote by $x = (x_1, ..., x_n)$ a point in euclidean n-space E^n. Let D be a domain in the u-plane, $u = (u_1, u_2)$. We shall define provisionally a *surface in E^n* to be a differentiable transformation $x(u)$ of some domain D into E^n. Later on (in §6) we shall give a global definition of a surface in E^n, but until then we shall use the word "surface" in the above sense.

Let us denote the *Jacobian matrix* of the mapping $x(u)$ by

$$M = (m_{ij}); \quad m_{ij} = \frac{\partial x_i}{\partial u_j}, \quad i = 1, ..., n; \quad j = 1, 2.$$

We note that the columns of M are the vectors

$$\frac{\partial x}{\partial u_j} = \left(\frac{\partial x_1}{\partial u_j}, ..., \frac{\partial x_n}{\partial u_j} \right)$$

For two vectors $v = (v_1, ..., v_n)$, $w = (w_1, ..., w_n)$, we denote the *inner product* by

$$v \cdot w = \sum_{k=1}^{n} v_k w_k ,$$

and the *exterior product* by

$$v \wedge w; \quad v \wedge w \; \epsilon \; E^N, \quad N = \binom{n}{2} ,$$

where the components of $v \wedge w$ are the determinants

$$\det \begin{pmatrix} v_i & v_j \\ w_i & w_j \end{pmatrix} , \quad i < j ,$$

arranged in some fixed order.

Finally, let us introduce the matrix

(1.1) $G = (g_{ij}) = M^{\top}M; \quad g_{ij} = \sum_{k=1}^{n} \frac{\partial x_k}{\partial u_i} \frac{\partial x_k}{\partial u_j} = \frac{\partial x}{\partial u_i} \cdot \frac{\partial x}{\partial u_j} \, ,$

and let us recall the identity of Lagrange:

(1.2) $\det G = \left| \frac{\partial x}{\partial u_1} \wedge \frac{\partial x}{\partial u_2} \right|^2 = \sum_{1 \le i < j \le n} \left(\frac{\partial(x_i, x_j)}{\partial(u_1, u_2)} \right)^2 .$

We may now formulate the following elementary lemma, which is of a purely algebraic nature.

LEMMA 1.1. *Let $x(u)$ be a differentiable map: $D \to E^n$. At each point of D the following conditions are equivalent:*

(1.3)   *the vectors $\dfrac{\partial x}{\partial u_1}$, $\dfrac{\partial x}{\partial u_2}$ are independent,*

(1.4)   *the Jacobian matrix M has rank 2,*

(1.5) $\exists \ i, j: \ 1 \le i < j \le n,$ *such that* $\dfrac{\partial(x_i, x_j)}{\partial(u_1, u_2)} \ne 0 ,$

(1.6) $\dfrac{\partial x}{\partial u_1} \wedge \dfrac{\partial x}{\partial u_2} \ne 0,$

(1.7) $\det G > 0 .$

Proof: Formula (1.2), combined with elementary properties of the rank of a matrix, gives the equivalence. ♦

DEFINITION. A surface S is *regular* at a point if the conditions of Lemma 1.1 hold at that point; S is *regular* if it is regular at every point of D.

We shall write $S \in C^r$ if $x(u) \in C^r$ in D; i.e., each coordinate x_k is an r-times continuously differentiable function of u_1, u_2 in D.

We shall assume throughout that $S \in C^r$, $r \geq 1$.

Suppose that S is a surface $x(u) \in C^r$ in D, and that $u(\tilde{u}) \in C^r$ is a diffeomorphism of a domain $\tilde{D}$ onto D. We shall say that the surface $\tilde{S}$ defined by $x(u(\tilde{u}))$ in $\tilde{D}$ *is obtained from S by a change of parameter.* We say that a property of S is *independent of parameters* if it holds at corresponding points of all surfaces $\tilde{S}$ obtained from S by a change of parameter. It is the object of differential geometry to study precisely those properties which are independent of parameters. Let us give some examples.

We note first that if the Jacobian matrix of the transformation $u(\tilde{u})$ is

$$U = (u_{ij}): \quad u_{ij} = \frac{\partial u_i}{\partial \tilde{u}_j} \quad ,$$

then the fact that $u(\tilde{u})$ is a diffeomorphism implies that

$$\frac{\partial(u_1, u_2)}{\partial(\tilde{u}_1, \tilde{u}_2)} = \det U \neq 0 \text{ in } \tilde{D} .$$

Furthermore, by the chain rule, it follows from $S \in C^r$ and $u(\tilde{u}) \in C^r$ that $\tilde{S} \in C^r$, so that the property of belonging to C^r is independent of (C^r changes of) parameters. In particular,

$$\frac{\partial x_i}{\partial \tilde{u}_k} = \sum_{j=1}^{2} \frac{\partial x_i}{\partial u_j} \frac{\partial u_j}{\partial \tilde{u}_k} \quad ,$$

or

$$\tilde{M} = MU,$$

whence

(1.8) $$\tilde{G} = U^{\mathsf{T}} G U$$

and

$$(1.9) \qquad \det \tilde{G} = \det G (\det U)^2 = \det G \left(\frac{\partial(u_1, u_2)}{\partial(\tilde{u}_1, \tilde{u}_2)} \right)^2 .$$

An immediate consequence of this equation, in view of (1.7) is that the property of S being regular at a point is independent of parameters.

Suppose now that Δ is a subdomain of D such that $\overline{\Delta} \subset D$, where $\overline{\Delta}$ is the closure of Δ. Let Σ be the restriction of the surface $x(u)$ to $u \in \Delta$. We define the *area of* Σ to be

$$(1.10) \qquad A(\Sigma) = \iint_{\Delta} \sqrt{\det G} \, du_1 \, du_2 .$$

If $u(\tilde{u})$ is a change of parameter, and $\tilde{\Delta}$ maps onto Δ, then the corresponding surface $\tilde{\Sigma}$ has area

$$A(\tilde{\Sigma}) = \iint_{\tilde{A}} \sqrt{\det \tilde{G}} \, d\tilde{u}_1 \, d\tilde{u}_2 = \iint_{\tilde{\Delta}} \sqrt{\det G} \left| \frac{\partial(u_1, u_2)}{\partial(\tilde{u}_1, \tilde{u}_2)} \right| d\tilde{u}_1 \, d\tilde{u}_2$$

$$= \iint_{\Delta} \sqrt{\det G} \, du_1 \, du_2 = A(\Sigma)$$

using (1.9) and the rule for change of variable in a double integral. Thus the area of a surface is independent of parameters.

We next note a special choice of parameters which is often useful to consider. Let i and j denote any two fixed distinct integers from 1 to n, and let D be a domain in the x_i, x_j plane. The equations

$$(1.11) \qquad x_k = f_k(x_i, x_j), \quad k = 1, \ldots, n; \quad k \neq i, j; \quad (x_i, x_j) \in D$$

define a surface S in E^n. A surface defined in this way will be said to be given in *nonparametric* or *explicit form*. This is, of course, a special case of the surfaces we have been considering up to now, the parameters being chosen to be two of the coordinates in E^n. In other words, we may rewrite (1.11) in the form

$$(1.12) \qquad x_i = u_1, \; x_j = u_2, \; x_k = f_k(u_1, u_2), \quad k \neq i, j.$$

In the classical case $n = 3$ we have a single function f_k, and the surface is defined by expressing one of the coordinates as a function of the other two.

In order for a surface to be expressible in non-parametric form, it is of course necessary for the projection map

$$(1.13) \qquad (x_1, \ldots, x_n) \to (x_i, x_j)$$

when restricted to the surface, to be one-to-one. This is not true in general for the whole surface, but we have the following important lemma.

LEMMA 1.2. *Let S be a surface $x(u)$, and let $u = a$ be a point at which S is regular. Then there exists a neighborhood Δ of a, such that the surface Σ obtained by restricting $x(u)$ to Δ has a reparametrization $\tilde{\Sigma}$ in non-parametric form.*

Proof: By condition (1.5) for regularity, and using the inverse mapping theorem, we deduce that there exists a neighborhood Δ of a in which the map $(u_1, u_2) \to (x_i, x_j)$ is a diffeomorphism. Furthermore, if $x(u) \in C^r$, the inverse map $(x_i, x_j) \to (u_1, u_2)$ is also C^r, and the same is true of the composed map

$$(1.14) \qquad (x_i, x_j) \to (u_1, u_2) \to (x_1, \ldots, x_n)$$

which defines $\tilde{\Sigma}$. ♦

Thus, when studying the local behavior of a surface, we may, whenever it is convenient, assume that the surface is in non-parametric form. Let us note also that the reparametrization (1.14) shows that in a neighborhood of a regular point the mapping $x(u)$ is always one-to-one.

In order to study more closely the behavior of a surface near a given point, we consider the totality of curves passing through the point and lying on the surface. First of all, by a curve C in E^n we shall mean a continuously differentiable map

$$(1.15) \qquad\qquad \phi : [\alpha, \beta] \to E^n$$

where $[\alpha, \beta]$ is some interval on the real line. We shall also use the notation

$$(1.16) \qquad x = \phi(t), \quad \alpha \leq t \leq \beta ; \quad \phi(t) = (\phi_1(t), ..., \phi_n(t)) \, \epsilon \, C^1 .$$

The *tangent vector* to the curve at a point t_0 is the vector

$$(1.17) \qquad\qquad x'(t_0) = (\phi_1'(t_0), ..., \phi_n'(t_0)).$$

The curve is *regular* at t_0 if $x'(t_0) \neq 0$.

Suppose now that we have a surface S defined by $x(u)$, $u \, \epsilon \, D$, and a curve C defined by (1.15). We shall say that C *lies on* S, if the image of $[\alpha, \beta]$ under ϕ is included in the image of D under $x(u)$. Since we are interested now in the local study of S, let us choose any point $u = a$ at which S is regular, and let us restrict $x(a)$ to a neighborhood of a in which Lemma 1.2 is valid. We shall continue to denote this restricted domain by D, and the surface by S. Then we have the representation (1.14) and the fact that the mapping $x(u)$ is one-to-one in D. We consider the totality of curves C which lie on S and pass through the point $b = x(a)$.

In order to fix the notation, we may assume that there is a fixed value t_0, $\alpha < t_0 < \beta$, such that for each curve C, $\phi(t_0) = b$. By virtue of the representation (1.14), to each such curve C corresponds a curve $u(t)$ in D such that $u(t_0) = a$. Conversely, to each curve $u(t)$ in D with $u(t_0) = a$ corresponds obviously a curve $\phi(t) = x(u(t))$ on S, with $\phi(t_0) = b$. For the tangent vector to the curve C we have the formula

$$(1.18) \qquad x'(t_0) = u_1'(t_0)\frac{\partial x}{\partial u_1} + u_2'(t_0)\frac{\partial x}{\partial u_2} \;,$$

where $\partial x/\partial u_1$ and $\partial x/\partial u_2$ are evaluated at $u = a$.

LEMMA 1.3. *At a regular point of a surface S, if we consider the set of all curves which lie on S and pass through the point, then the totality of their tangent vectors at the point form a two-dimensional vector space.*

Proof: Since we can obviously find curves $u(t)$ in D such that $u(t_0) = a$, and $u_1'(t_0)$, $u_2'(t_0)$ take on arbitrarily assigned values, it follows from (1.18) that the set of tangent vectors $x'(t_0)$ consists of all linear combinations of the two vectors $\partial x/\partial u_1$ and $\partial x/\partial u_2$. But by condition (1.3) for regularity these vectors are independent and therefore span a two-dimensional space. ♦

DEFINITION. The vector space described in Lemma 1.3 is called the *tangent plane* to the surface S at the point $b = x(a)$, and is denoted by Π or $\Pi(a)$.

Thus a surface S has at every regular point a tangent plane, which by its definition is independent of parameters.

For the length of a tangent vector we have from (1.1) and (1.18):

$$(1.19) \qquad |x'(t_0)|^2 = x'(t_0) \cdot x'(t_0) = \sum_{i,j=1}^{2} g_{ij} u_i'(t_0) u_j'(t_0) \, .$$

Thus, the square of the length is expressed by a quadratic form in the corresponding tangent vector $u'(t_0)$, with matrix G; this quadratic form is often referred to as the *first fundamental form* of the surface S. We have seen in (1.10) that the determinant of this form defines areas on the surface. Similarly, the length of curves on the surface are obtained from (1.19), since the length of the curve $x(t)$, $a \le t \le \beta$ in E^n is given by

$$(1.20) \qquad L = \int_a^\beta |x'(t)| dt \, .$$

It is convenient to associate with an arbitrary curve C of the form (1.16) the quantity

$$(1.21) \qquad s(t_0) = \int_a^{t_0} |x'(t)| dt \, .$$

Since $s'(t_0) = |x'(t_0)| \ge 0$ for $a \le t_0 \le \beta$, we have a monotone mapping

$$(1.22) \qquad s(t): [a, \beta] \to [0, L] \, .$$

If furthermore, the curve C is regular, then $s'(t) = |x'(t)| > 0$, and the map (1.22) has a differentiable inverse $t(s)$. The composed map

$$(1.23) \qquad \tilde{\phi}(s): [0, L] \xrightarrow{t(s)} [a, \beta] \xrightarrow{\phi(t)} E^n$$

defines a curve $\tilde{C}$ which is called the *parametrization of C with respect to arclength*. We have at each point the *unit tangent vector*

(1.24) $\qquad T = \dfrac{dx}{ds} = \dfrac{x'(t)}{s'(t)} \ ; \ \left|\dfrac{dx}{ds}\right| = \dfrac{|x'(t)|}{s'(t)} = 1 \ .$

We wish next to study "second order effects," and we shall assume from now on that all curves considered are regular C^2 curves. We may then introduce the parametrization (1.23) with respect to arc length, and we define at each point the *curvature vector*

(1.25) $\qquad \dfrac{d^2x}{ds^2} = \dfrac{dT}{ds} \ ,$

as the derivative of the unit tangent with respect to arc length. We use the same notation as in the paragraph preceding Lemma 1.3, but we now add the assumption that the surface $S \in C^2$, and we restrict the class of curves passing through the regular point $b = x(a)$ on S to regular C^2-curves lying on S. We shall seek to describe the totality of curvature vectors to these curves evaluated at the point $b = x(a)$. More precisely, if Π is the tangent plane to S at this point, let us denote by $\Pi^{\perp}$ its orthogonal complement, an $(n-2)$-dimensional space called the *normal space* to S at the point. Each vector is determined by its projections in Π and $\Pi^{\perp}$. For our present purposes we shall examine the projection of the curvature vector into $\Pi^{\perp}$.

An arbitrary vector $N \in \Pi^{\perp}$ is called a *normal* to S. Since such a vector is in particular orthogonal to $\partial x/\partial u_1$, $\partial x/\partial u_2$, we may compute as follows:

$$\dfrac{dx}{ds} = \sum_i \dfrac{du_i}{ds} \dfrac{\partial x}{\partial u_i} \ , \qquad \dfrac{d^2x}{ds^2} = \sum_i \dfrac{d^2u_i}{ds^2}\left(\dfrac{\partial x}{\partial u_i}\right) + \sum_{i,j} \dfrac{du_i}{ds}\dfrac{du_j}{ds}\dfrac{\partial^2 x}{\partial u_i \partial u_j}$$

(1.26)

$$\dfrac{d^2x}{ds^2} \cdot N = \sum b_{ij}(N) \dfrac{du_i}{ds}\dfrac{du_j}{ds}$$

where we have introduced the notation

(1.27)
$$b_{ij}(N) = \frac{\partial^2 x}{\partial u_i \partial u_j} \cdot N \, ,$$

the vector $\partial^2 x/\partial u_i \partial u_j$ being evaluated at the point $u = a$. By noting that

$$\left(\frac{ds}{dt}\right)^2 = |x'(t)|^2 = \sum g_{ij} u_i'(t) u_j'(t)$$

and that $du_i/ds = (du_i/dt)/(ds/dt)$, we may rewrite (1.26) in the form

$$k = \frac{II}{I}$$

(1.28)
$$\frac{d^2 x}{ds^2} \cdot N = \frac{\sum b_{ij}(N) u_i'(t_0) u_j'(t_0)}{\sum g_{ij} u_i'(t_0) u_j'(t_0)}$$

$$\frac{II}{I}$$

The numerator on the right hand side is a quadratic form in the tangent vector $u'(t_0)$ whose matrix $b_{ij}(N)$ depends on the point of the surface and the normal N. It is called the *second fundamental form* of S with respect to N. We note that the entire right-hand side of (1.28) depends on the particular curve C only to the extent of the tangent vector to C at the point. In fact, the homogeneity of the right-hand side in the components of $u'(t_0)$ shows that it depends only on the *direction* of the tangent vector: i.e., on the unit tangent T. We may therefore rewrite (1.28) in the form

(1.29)
$$\frac{d^2 x}{ds^2} \cdot N = k(N, T), \quad N \, \epsilon \, \Pi^{\perp}, \quad T \, \epsilon \, \Pi,$$

where the right-hand side is at each point of S a well-defined function of the normal N and the unit tangent T, called the *normal curvature* of S in the direction T with respect to the normal N. If we fix N, and let T vary, we obtain the quantities

(1.30) $k_1(N) = \max\limits_{T} k(N, T), \quad k_2(N) = \min\limits_{T} k(N, T)$

called the *principal curvatures* of S at the point, with respect to
the normal N. Finally we introduce the average value

(1.31) $$H(N) = \frac{k_1(N) + k_2(N)}{2},$$

called the *mean curvature of* S at the point, with respect to the
normal N.

To obtain an explicit expression for $H(N)$, we note that since
the right-hand side of (1.28) is the quotient of quadratic forms, its
maximum and minimum, which we have denoted by $k_1(N)$, $k_2(N)$, are
the roots of the equation

(1.32) $\det(b_{ij}(N) - \lambda g_{ij}) = 0 .$

Expanded, this equation takes the form

$\det(g_{ij})\lambda^2 - (g_{22}b_{11}(N) + g_{11}b_{22}(N) - 2g_{12}b_{12}(N))\lambda + \det(b_{ij}(N)) = 0.$

For the sum of the roots, we therefore have

(1.33) $$H(N) = \frac{g_{22}b_{11}(N) + g_{11}b_{22}(N) - 2g_{12}b_{12}(N)}{2\det(g_{ij})} .$$

It follows immediately from the definition (1.27) that the $b_{ij}(N)$
are linear in N, and from (1.33) that $H(N)$ is linear in N for $N \in \Pi^{\perp}$.
Thus there exists a unique vector $H \in \Pi^{\perp}$ such that

(1.34) $H(N) = H \cdot N$ for all $N \in \Pi^{\perp} .$

The vector H thus defined is called the *mean curvature vector* of
S at the point. If $e_1, \ldots, e_{n-2}$ is any orthonormal basis of $\Pi^{\perp}$, it
follows from (1.34) that the mean curvature vector H may be

E F G

expressed as   *L m N*

(1.35)
$$H = \sum_{k=1}^{n-2} [H(e_k)]\, e_k \ .$$

DEFINITION. A surface S is a *minimal surface* if its mean curvature vector H vanishes at every point.

The reason for this terminology will become apparent in Section 3. For the moment we note merely that by virtue of (1.34) and (1.35), $H = 0$ if and only if $H(N) = 0$ for all $N \in \Pi^{\perp}$. Thus, using (1.33), minimal surfaces are characterized in terms of their first and second fundamental forms by the equation

(1.36) $g_{22}b_{11}(N) + g_{11}b_{22}(N) - 2g_{12}b_{12}(N) = 0.$

E L + G N -2 F m = 0

pg 65 Strick

⟹   *A du² + 2B du dv + c dv² = 0*
          orthogonal net

A skew-symmul
⊘A orgonal .

Pg6

§2. *Non-parametric surfaces.*

We consider in this section surfaces in the non-parametric form (1.11). By relabeling the coordinates in E^n we may assume that the surface is defined by

unit vector
Flanders
pg 126

$$(2.1) \qquad x_k = f_k(x_1, x_2), \qquad k = 3, ..., n ;$$

or equivalently

$$(2.2) \qquad x_1 = u_1, \quad x_2 = u_2, \quad x_k = f_k(u_1, u_2), \quad k = 3, ..., n .$$

Then

$$(2.3) \quad \frac{\partial x}{\partial u_1} = \left(1, 0, \frac{\partial f_3}{\partial u_1}, ..., \frac{\partial f_n}{\partial u_1}\right), \quad \frac{\partial x}{\partial u_2} = \left(0, 1, \frac{\partial f_3}{\partial u_2}, ..., \frac{\partial f_n}{\partial u_2}\right) ,$$

and

Pg 24

$$(2.4) \qquad g_{11} = 1 + \sum_{k=3}^{n} \left(\frac{\partial f_k}{\partial u_1}\right)^2 , \qquad g_{12} = \sum_{k=3}^{n} \frac{\partial f_k}{\partial u_1} \frac{\partial f_k}{\partial u_2} ,$$

$$g_{22} = 1 + \sum_{k=3}^{n} \left(\frac{\partial f_k}{\partial u_2}\right)^2 .$$

We note that the vectors $\partial x/\partial u_1$, $\partial x/\partial u_2$ are obviously independent, so that every surface in non-parametric form is automatically regular.

We again denote by Π the tangent plane, and by $\Pi^\perp$ the normal space.

LEMMA 2.1. *Let* $N_3, ..., N_n$ *be arbitrary. Then there exist unique* N_1, N_2 *such that the vector* $N = (N_1, ..., N_n)$ *is in* $\Pi^\perp$.

Proof: The vector N is in $\Pi^\perp$ if and only if $N \cdot \partial x/\partial u_i = 0$, $i = 1, 2$. By (2.3) we have $N_i = -\sum_{k=3}^{n} N_k(\partial f_k/\partial u_i)$, $i = 1, 2$. ♦

We give an application of this lemma to arbitrary surfaces.

LEMMA 2.2. *Let* $x(u)$ *define a surface* $S \in C^r$, *let* $b = x(a)$ *be a regular point of* S, *and let* N *be a normal to* S *at this point. Then there exists a neighborhood* Δ *of* a, *and* $N(u) \in C^{r-1}$ *in* Δ *such that* $N(u) \in \Pi^1(u)$ *and* $N(a) = N$.

Proof: By Lemma 1.2 we may find a neighborhood Δ of a in which S has a reparametrization in the form (2.1) (assuming suitable labeling of the coordinates in E^n). Let $N = (N_1, ..., N_n)$, and set

$$N_i(u) = - \sum_{k=3}^{n} N_k \frac{\partial f_k}{\partial u_i} \quad , \qquad i = 1, 2.$$

Then $N(u)$ has the desired properties. ◆

For the remainder of this section we consider surfaces $S \in C^2$. From (2.3) we deduce

$$(2.5) \qquad \frac{\partial^2 x}{\partial u_i \partial u_j} = \left(0, 0, \frac{\partial^2 f_3}{\partial u_i \partial u_j}, ..., \frac{\partial^2 f_n}{\partial u_i \partial u_j} \right) \quad .$$

Thus, for an arbitrary normal $N = (N_1, ..., N_n)$, we have

$$(2.6) \qquad b_{ij}(N) = \sum_{k=3}^{n} N_k \frac{\partial^2 f_k}{\partial u_i \partial u_j} \quad .$$

The equation (1.36) for a minimal surface therefore takes the form

$$\sum_{k=3}^{n} \left[\left(1 + \sum_{m=3}^{n} \left(\frac{\partial f_m}{\partial u_2} \right)^2 \right) \frac{\partial^2 f_k}{\partial u_1^2} - 2 \left(\sum_{m=3}^{n} \frac{\partial f_m}{\partial u_1} \frac{\partial f_m}{\partial u_2} \right) \frac{\partial^2 f_k}{\partial u_1 \partial u_2} \right.$$

$$\left. + \left(1 + \sum_{m=3}^{n} \left(\frac{\partial f_m}{\partial u_1} \right)^2 \right) \frac{\partial^2 f_k}{\partial u_2^2} \right] N_k = 0,$$

for all normal vectors N. Since, as we observed earlier, the components $N_3, ..., N_n$ may be chosen arbitrarily, it follows that each of the coefficients of N_k must vanish, for $k = 3, ..., n$. We thus obtain $n - 2$ equations for the $n - 2$ functions $f_3, ..., f_n$. If we recall that $u_1 = x_1$, $u_2 = x_2$, and if we introduce the vector notation

$$(2.7) \qquad f(x_1, x_2) = (f_3(x_1, x_2), ..., f_n(x_1, x_2)) ,$$

then these equations may be written as a single vector equation

$$(2.8) \quad \left(1 + |\frac{\partial f}{\partial x_2}|^2\right)\frac{\partial^2 f}{\partial x_1^2} - 2\left(\frac{\partial f}{\partial x_1} \cdot \frac{\partial f}{\partial x_2}\right)\frac{\partial^2 f}{\partial x_1 \partial x_2}$$

$$+ \left(1 + |\frac{\partial f}{\partial x_1}|^2\right)\frac{\partial^2 f}{\partial x_2^2} = 0.$$

This is the *minimal surface equation* for non-parametric minimal surfaces in E^n. Every regular minimal surface provides local solutions of this equation, by Lemma 1.2. We shall use this fact later on to aid in the local study of minimal surfaces. For the present, let us observe that Equation (2.8) allows us to find explicitly a number of specific examples of minimal surfaces. Some of these surfaces turn out to be very useful in the general theory because of certain extremal properties which they possess.

First let us consider the case $n = 3$. Then $f(x_1, x_2) = f_3(x_1, x_2)$ and (2.8) reduces to a single equation for the scalar function $f(x_1, x_2)$. We have the following classical surfaces.

The helicoid:

$$(2.9) \qquad f(x_1, x_2) = \tan^{-1}\frac{x_2}{x_1} .$$

One can show that this is the only solution of (2.8) which is also a harmonic function, and that the helicoid is the only ruled minimal surface. (For a more detailed discussion of all these surfaces we refer the reader to Darboux [1].)

Pg88 Struik

$$y = a \cosh^{-1} \frac{x}{a}$$

The catenoid:

(2.10) $f(x_1, x_2) = \cosh^{-1} r, \quad r = \sqrt{x_1^2 + x_2^2}$,

or

(2.11) $x_1^2 + x_2^2 = (\cosh x_3)^2$.

This is the only minimal surface which is also a surface of rotation.

Scherk's surface:

(2.12) $f(x_1, x_2) = \log \dfrac{\cos x_2}{\cos x_1}$.

This is the only minimal surface of translation; that is (2.12) is the only solution of (2.8) of the form $f(x_1, x_2) = g(x_1) + h(x_2)$.

There are two remarks we should make about these solutions. First of all, the image of a minimal surface under a similarity transformation is again a minimal surface, so that one can obtain other solutions of (2.8) trivially by this method. Second of all, we have not specified the domain of definition of the above solutions, but we may note that none of them is defined for all x_1, x_2. This turns out not to be an accident, since the theorem of Bernstein which we shall prove later on (Section 5) states that for $n = 3$ there are no non-trivial solutions of Equation (2.8) valid in the whole x_1, x_2-plane.

We turn now to the case of arbitrary n. We note first that if each f_k is linear in x_1, x_2, then (2.8) is satisfied trivially. In this case the surface S is a plane.

If n is even we have the following special solutions. Let $z = x_1 + ix_2$, and let $g_1(z), \ldots, g_m(z)$ be complex analytic functions of z, where $n = 2m + 2$. Then setting

$$f_k(x_1, x_2) = \begin{cases} \text{Re}\{g_p(z)\}, & k = 2p + 1 \\ \text{Im}\{g_p(z)\}, & k = 2p + 2 \end{cases}$$

for $p = 1, \ldots, m$, we obtain by direct verification a solution of (2.8). We may word this result as follows: *the graph of a complex-analytic curve, considered as a surface in real euclidean space, is always a minimal surface.* For the case of a single function $g_1(z)$, the corresponding minimal surfaces in E^4 were studied in great detail by Kommerell [1].

Scherk's Surface

$$F(z) = \frac{2}{(1 - z^4)}''$$

Cartesian

$$(\cos x)e^z = \cos y$$

§3. *Surfaces that minimize area.*

We shall now discuss briefly the problem which led historically to the theory of minimal surfaces. Namely, that of characterizing those surfaces which have least area among all surfaces with the same boundary. Specifically, we shall consider the following situation.

Let S be a regular surface defined by $x(u) \in C^2$ in a domain D. Let Γ be a closed curve in D which bounds a subdomain Δ, and let Σ be the surface defined by $x(u)$ restricted to Δ. Suppose that the area of Σ is less than or equal to the area of every surface $\tilde{\Sigma}$ defined by $\tilde{x}(u)$ in Δ such that for u on Γ, $\tilde{x}(u) = x(u)$. What does this imply about the surface $x(u)$?

We shall apply the standard methods of the calculus of variations in two different forms. First we shall make normal variations of the surface, and later we shall consider non-parametric surfaces and variations perpendicular to the x_1, x_2-plane.

To start with, let us suppose that $N(u) \in C^1$ in D, such that $N(u)$ is normal to S at $x(u)$. That is

$$(3.1) \qquad N(u) \cdot \frac{\partial x}{\partial u_i} \equiv 0, \qquad i = 1, 2.$$

Differentiating this equation yields

$$(3.2) \qquad \frac{\partial N}{\partial u_j} \cdot \frac{\partial x}{\partial u_i} = -N \cdot \frac{\partial^2 x}{\partial u_i \partial u_j} = -b_{ij}(N) \ .$$

We now consider an arbitrary function $h(u) \in C^2$ in D, and for each real number λ we form the surface

$$S_\lambda: \quad \tilde{x}(u) = x(u) + \lambda h(u) N(u), \qquad u \in D.$$

We find

$$\frac{\partial \tilde{x}}{\partial u_i} = \frac{\partial x}{\partial u_i} + \lambda \left[h \frac{\partial N}{\partial u_i} + \frac{\partial h}{\partial u_i} N \right] ,$$

and using (3.1), (3.2)

$$\tilde{g}_{ij} = \frac{\partial \tilde{x}}{\partial \tilde{u}_i} \cdot \frac{\partial \tilde{x}}{\partial u_j} = g_{ij} - 2\lambda h b_{ij}(N) + \lambda^2 c_{ij}$$

where c_{ij} is a continuous function of u in D.

It follows that

(3.3) $$\det \tilde{g}_{ij} = a_0 + a_1 \lambda + a_2 \lambda^2 ,$$

where

(3.4) $a_0 = \det g_{ij}, \quad a_1 = -2h(g_{11} b_{22}(N) + g_{22} b_{11}(N) - 2 g_{12} b_{12}(N))$

and a_2 is a continuous function of u_1, u_2, λ for u in D.

As a first consequence of this formula, using the fact that S is regular we deduce that a_0 has a positive minimum on $\overline{\Delta}$, and since a_1 and a_2 are continuous in D, there exists $\varepsilon > 0$ such that $\det \tilde{g}_{ij} > 0$ for $|\lambda| < \varepsilon$ and $u \in \overline{\Delta}$. In other words, for $|\lambda| < \varepsilon$, the surfaces Σ_λ defined by restricting $\tilde{x}(u)$ to Δ are all regular sur-faces. To compute their area $A(\lambda) = A(\Sigma_\lambda)$, we note that in view of (3.3), we have

(3.5) $$\left| \sqrt{\det \tilde{g}_{ij}} - \left(\sqrt{a_0} + \frac{a_1}{2\sqrt{a_0}} \lambda \right) \right| < M \lambda^2$$

for u in Δ, where M is a positive constant. By (1.10) and (3.4) we have

$$A(0) = A(\Sigma) = \iint_\Delta \sqrt{a_0} \, du_1 \, du_2$$

and integrating (3.5)

$$\left| A(\lambda) - A(0) - \lambda \iint_\Delta \frac{a_1}{2\sqrt{a_0}}\, du_1\, du_2 \right| < M_1 \lambda^2 \ ,$$

or

$$\left| \frac{A(\lambda) - A(0)}{\lambda} - \iint_\Delta \frac{a_1}{2\sqrt{a_0}}\, du_1\, du_2 \right| < M_1 \lambda \ .$$

Letting λ tend to zero, substituting in the expressions (3.4) for a_0, a_1, and recalling the formula (1.33) for the mean curvature, we arrive at the expression Pg 13

(3.6)
$$A'(0) = -2 \iint_\Delta H(N)\, h(u) \sqrt{\det g_{ij}}\, du_1\, du_2$$

for the rate of change of area as a function of λ.

We may note in passing that if $f(u)$ is an arbitrary continuous function of u in $\overline{\Delta}$ we may define the *integral of f with respect to surface area on* Σ as

(3.7)
$$\iint_\Sigma f(u)\, dA = \iint_\Delta f(u)\sqrt{\det g_{ij}}\, du_1\, du_2 \ .$$

With this notation, if we choose our family of surfaces S_λ by setting $h(u) \equiv 1$, then formula (3.6) reduces to

$$A'(0) = -2 \iint_\Sigma H(N)\, dA$$

which provides an interesting interpretation of the quantity $H(N)$.

We now return to our original problem, and we make the following assertion: *in order for* Σ *to minimize area, its mean curvature must be identically zero.* This follows immediately from (3.6) using the standard argument of the calculus of variations. Namely, if the mean curvature did not vanish identically, then there would be a point $u = a$ in Δ and a normal $N = N(a)$ such that $H(N) \neq 0$. We may assume $H(N) > 0$. By Lemma 2.2 we can find a neighborhood V_1 of a, and $N(u) \in C^1$ in V_1, such that $N(u)$ is normal to S at $x(u)$. Then we will have $H(N) > 0$ throughout a neighborhood V_2, $a \in V_2 \subset V_1$, and if we choose the function $h(u)$ so that $h(a) > 0$, $h(u) \geq 0$ for all u, and $h(u) \equiv 0$ for $u \notin V_2$, the integral on the right-hand side of (3.6) will be strictly positive. However, if V_2 is small enough so that $V_2 \subset \Delta$, then $\tilde{x}(u) = x(u)$ on Γ, so that Σ_λ will be a surface with the same boundary as Σ. The assumption that Σ minimizes area implies that $A(\lambda) \geq A(0)$ for all λ, whence $A'(0) = 0$. Thus we would obtain a contradiction to (3.6), and the assertion is proved.

Thus minimal surfaces arose originally in connection with minimizing area, and it is from this connection that they derived their name. However, as we shall see, they also arise naturally in a number of other connections, and many of their most important properties are totally unrelated to questions of area. Before leaving the subject we shall use the property of minimizing area to derive several other forms of the minimal surface equation.

We start from a surface in non-parametric form:

$$x_k = f_k(x_1, x_2), \qquad k = 3, \dots, n,$$

and introduce the vector notation

$$(3.8) \qquad f = (f_3, \dots, f_n), \quad p = \frac{\partial f}{\partial x_1}, \quad q = \frac{\partial f}{\partial x_2}, \quad r = \frac{\partial^2 f}{\partial x_1^2},$$

$$r = \frac{\partial^2 f}{\partial x_1^2}, \quad s = \frac{\partial^2 f}{\partial x_1 \partial x_2}, \quad t = \frac{\partial^2 f}{\partial x_2^2}$$

Then the minimal surface equation (2.8) may be written as

$$(3.9) \quad (1 + |q|^2)\frac{\partial p}{\partial x_1} - (p \cdot q)\left(\frac{\partial p}{\partial x_2} + \frac{\partial q}{\partial x_1}\right) + (1 + |p|^2)\frac{\partial q}{\partial x_2} = 0,$$

or as

$$(3.10) \quad (1 + |q|^2)r - 2(p \cdot q)s + (1 + |p|^2)t = 0 .$$

The equations (2.4) take the form

$$(3.11) \quad g_{11} = 1 + |p|^2, \quad g_{12} = p \cdot q, \quad g_{22} = 1 + |q|^2,$$

whence

$$(3.12) \quad \det g_{ij} = 1 + |p|^2 + |q|^2 + |p|^2|q|^2 - (p \cdot q)^2 .$$

One often uses the notation

$$(3.13) \quad W = \sqrt{\det g_{ij}}$$

for non-parametric surfaces.

Suppose now that we make a variation in our surface, setting

$$\tilde{f}_k = f_k + \lambda h_k, \quad k = 3, ..., n ,$$

where λ is a real number, and $h_k \in C^1$ in the domain of definition D of the f_k. In vector notation, setting $h = (h_3, ..., h_n)$ we have

$$\tilde{f} = f + \lambda h, \quad \tilde{p} = p + \lambda\frac{\partial h}{\partial x_1}, \quad \tilde{q} = q + \lambda\frac{\partial h}{\partial x_2} ,$$

whence

$$\tilde{W}^2 = W^2 + 2\lambda X + \lambda^2 Y ,$$

where

$$X = [(1 + |q|^2)p - (p \cdot q)q] \cdot \frac{\partial h}{\partial x_1} + [(1 + |p|^2)q - (p \cdot q)p] \cdot \frac{\partial h}{\partial x_2}$$

and Y is continuous in x_1, x_2. It follows that

$$\tilde{W} = W + \lambda \frac{X}{W} + \lambda^2 Z$$

where Z is again continuous.

We now consider a closed curve Γ in the domain of definition of $f(x_1, x_2)$, and let Δ be the region bounded by Γ. If the surface $x_k = f(x_1, x_2)$ over Δ minimizes the area among all surfaces with the same boundary, then for every choice of h such that $h = 0$ on Γ, we have

$$\iint_\Delta \tilde{W} \, dx_1 \, dx_2 \geq \iint_\Delta W \, dx_1 \, dx_2$$

which is only possible if

$$\iint_\Delta \frac{X}{W} = 0 \ .$$

Substituting in the above expression for X, integrating by parts, and using the fact that $h = 0$ on Γ, we find

$$\iint_\Delta \left[\frac{\partial}{\partial x_1} \left[\frac{1 + |q|^2}{W} p - \frac{p \cdot q}{W} q \right] + \frac{\partial}{\partial x_2} \left[\frac{1 + |p|^2}{W} q - \frac{p \cdot q}{W} p \right] \right] h \, dx_1 \, dx_2 = 0.$$

By the same reasoning as above, it follows that the equation

$$(3.14) \quad \frac{\partial}{\partial x_1} \left[\frac{1 + |q|^2}{W} p - \frac{p \cdot q}{W} q \right] + \frac{\partial}{\partial x_2} \left[\frac{1 + |p|^2}{W} q - \frac{p \cdot q}{W} p \right] = 0$$

must hold everywhere.

Once we have found this equation, it is easy to verify that it is a consequence of the minimal surface equation (3.10). In fact the left-hand side of (3.14) may be written as the sum of three terms:

$$\left[\frac{1+|q|^2}{W}\frac{\partial p}{\partial x_1} - \frac{p\cdot q}{W}\left(\frac{\partial q}{\partial x_1} + \frac{\partial p}{\partial x_2}\right) + \frac{1+|p|^2}{W}\frac{\partial q}{\partial x_2}\right]$$

$$+\left[\frac{\partial}{\partial x_1}\left(\frac{1+|q|^2}{W}\right) - \frac{\partial}{\partial x_2}\left(\frac{p\cdot q}{W}\right)\right]p$$

$$+\left[\frac{\partial}{\partial x_2}\left(\frac{1+|p|^2}{W}\right) - \frac{\partial}{\partial x_1}\left(\frac{p\cdot q}{W}\right)\right]q\ .$$

The first term vanishes by (3.9). If we expand out the coefficient of p in the second term, we find the expression

(3.15) $\dfrac{\partial}{\partial x_1}\left(\dfrac{1+|q|^2}{W}\right) - \dfrac{\partial}{\partial x_2}\left(\dfrac{p\cdot q}{W}\right)$

$$= \frac{1}{W^3}[(p\cdot q)q - (1+|q|^2)p]\cdot[(1+|q|^2)r - 2(p\cdot q)s + (1+|p|^2)t]$$

which vanishes by (3.10). Interchanging p and q, x_1 and x_2, we see that the coefficient of q in the third term vanishes also, thus proving (3.14). In the process we have also shown that the two equations

(3.16)

$$\frac{\partial}{\partial x_1}\left(\frac{1+|q|^2}{W}\right) = \frac{\partial}{\partial x_2}\left(\frac{p\cdot q}{W}\right)$$

$$\frac{\partial}{\partial x_1}\left(\frac{p\cdot q}{W}\right) = \frac{\partial}{\partial x_2}\left(\frac{1+|p|^2}{W}\right)$$

are satisfied by every solution of the minimal surface equation (3.10). These equations have long been known in the case $n = 3$, and the fact that they are in divergence form allows one to derive many consequences which are not immediate from (3.10). (See, for example, Radó [3].) We shall see that Equations (3.16) have equally important consequences in the case of arbitrary n.

Equal Angles

§4. *Isothermal parameters.*

When studying properties of a surface which are independent of choice of parameters, it is convenient to choose parameters in such a way that geometric properties of the surface are reflected in the parameter plane. As an example one can ask that the mapping of the parameter plane onto the surface be conformal, so that angles between curves on the surface are equal to the angles between the corresponding curves in the parameter plane. Analytically, this condition is expressed in terms of the first fundamental form (1.1) by

$$(4.1) \qquad g_{11} = g_{22}, \quad g_{12} = 0$$

or

$$(4.2) \qquad g_{ij} = \lambda^2 \delta_{ij}, \quad \lambda = \lambda(u) > 0 .$$

Parameters u_1, u_2 satisfying these conditions are called *isothermal parameters*.

Many of the basic quantities considered in surface theory simplify considerably when referred to isothermal parameters. For example, from (4.2) we have

$$(4.3) \qquad \det g_{ij} = \lambda^4$$

pg 13

and the formula (1.33) for mean curvature becomes

$$(4.4) \qquad H(N) = \frac{b_{11}(N) + b_{22}(N)}{2\lambda^2} .$$

We also have the following useful formula for the Laplacian of the coordinate vector of an arbitrary surface.

LEMMA 4.1. *Let a regular surface S be defined by* $x(u) \in C^2$ *where* u_1, u_2 *are isothermal parameters. Then*

(4.5)
$$\Delta x = 2\lambda^2 H$$

where H is the mean curvature vector.

Proof: The defining equation (4.1) for isothermal parameters may be written in the form

$$\frac{\partial x}{\partial u_1} \cdot \frac{\partial x}{\partial u_1} = \frac{\partial x}{\partial u_2} \cdot \frac{\partial x}{\partial u_2} \quad , \quad \frac{\partial x}{\partial u_1} \cdot \frac{\partial x}{\partial u_2} = 0 \ .$$

Differentiating the first of these with respect to u_1, and the second with respect to u_2 yields

$$\frac{\partial^2 x}{\partial u_1^2} \cdot \frac{\partial x}{\partial u_1} = \frac{\partial^2 x}{\partial u_1 \partial u_2} \cdot \frac{\partial x}{\partial u_2} = - \frac{\partial^2 x}{\partial u_2^2} \cdot \frac{\partial x}{\partial u_1} \ ,$$

whence

$$\Delta x \cdot \frac{\partial x}{\partial u_1} = \left(\frac{\partial^2 x}{\partial u_1^2} + \frac{\partial^2 x}{\partial u_2^2} \right) \cdot \frac{\partial x}{\partial u_1} = 0 .$$

Similarly, differentiating the first equation with respect to u_2 and the second with respect to u_1 yields

$$\Delta x \cdot \frac{\partial x}{\partial u_2} = 0 \ .$$

Thus Δx is a vector perpendicular to the tangent plane to S. But if N is an arbitrary normal vector to S, we have

$$\Delta x \cdot N = \frac{\partial^2 x}{\partial u_1^2} \cdot N + \frac{\partial^2 x}{\partial u_2^2} \cdot N = b_{11}(N) + b_{22}(N) = 2\lambda^2 H(N)$$

by (4.4). This means that $\Delta x / 2\lambda^2$ is a normal vector which satisfies the defining equation (1.34) of the mean curvature vector H, and this proves (4.5). ◆

We may note that formula (4.5) is of interest in other connections, and in particular in the study of surfaces of constant mean curvature. However, for our purposes we are interested only in the following immediate consequence.

Harmonic

LEMMA 4.2. *Let* $x(u) \epsilon C^2$ *define a regular surface* S *in isothermal parameters. Necessary and sufficient that the coordinate functions* $x_k(u_1, u_2)$ *be* harmonic *is that* S *be a minimal surface.*

Thus we see that minimal surfaces arise naturally in quite a different context from that of minimizing area. We wish to pursue further the connection with harmonic functions.

Let us introduce the following notation. Given a surface $x(u)$, we consider the complex-valued functions

$$(4.6) \qquad \phi_k(\zeta) = \frac{\partial x_k}{\partial u_1} - i \frac{\partial x_k}{\partial u_2} \; ; \qquad \zeta = u_1 + iu_2 \; .$$

We note the identities:

$$\sum_{k=1}^{n} \phi_k^2(\zeta) = \sum_{k=1}^{n} \left(\frac{\partial x_k}{\partial u_1}\right)^2 - \sum_{k=1}^{n} \left(\frac{\partial x_k}{\partial u_2}\right)^2 - 2i \sum_{k=1}^{n} \frac{\partial x_k}{\partial u_1} \frac{\partial x_k}{\partial u_2}$$

$$(4.7) \qquad = \left| \frac{\partial x}{\partial u_1} \right|^2 - \left| \frac{\partial x}{\partial u_2} \right|^2 - 2i \frac{\partial x}{\partial u_1} \cdot \frac{\partial x}{\partial u_2} \; .$$

$$= g_{11} - g_{22} - 2ig_{12} \; .$$

$$(4.8) \qquad \sum_{k=1}^{n} |\phi_k(\zeta)|^2 = \sum_{k=1}^{n} \left(\frac{\partial x_k}{\partial u_1}\right)^2 + \sum_{k=1}^{n} \left(\frac{\partial x_k}{\partial u_2}\right)^2 = g_{11} + g_{22} \; .$$

We may read off directly the following properties of the functions $\phi_k(\zeta)$:

a) $\phi_k(\zeta)$ is analytic in $\zeta \Longleftrightarrow x_k$ is harmonic in u_1, u_2 ;

b) u_1, u_2 are isothermal parameters $\Longleftrightarrow$

$$(4.9) \qquad \sum_{k=1}^{n} \phi_k^2(\zeta) \equiv 0 ;$$

c) if u_1, u_2 are isothermal parameters, then S is regular $\Longleftrightarrow$

$$(4.10) \qquad \sum_{k=1}^{n} |\phi_k(\zeta)|^2 \neq 0 .$$

LEMMA 4.3. *Let* $x(u)$ *define a regular minimal surface, with* u_1, u_2 *isothermal parameters. Then the functions* $\phi_k(\zeta)$ *defined by* (4.6) *are analytic, and they satisfy equations* (4.9) *and* (4.10). *Conversely, let* $\phi_1(\zeta), ..., \phi_n(\zeta)$ *be analytic functions of* ζ *which satisfy* (4.9) *and* (4.10) *in a simply-connected domain D. Then there exists a regular minimal surface* $x(u)$ *defined over D, such that equations* (4.6) *are valid.*

Proof: The first statement follows immediately from properties a), b) and c), in view of Lemma 4.2. For the converse, if we define

$$(4.11) \qquad x_k = \text{Re} \int \phi_k(\zeta) d\zeta ,$$

then the x_k are harmonic functions satisfying (4.6), and again applying a), b), and c) in the opposite direction, the result follows from Lemma 4.2. ◆

Thus we see that the local study of regular minimal surfaces in E^n is equivalent to the study of n-tuples of analytic functions satisfying (4.9) and (4.10). We may note that by (4.11), these functions determine the x_k up to additive constants, and the surface is therefore determined up to a translation.

The preceding results are all based on the assumption that the surface can be represented locally in terms of isothermal parameters. However, the existence of such parameters is not at all obvious, and in the case of C^1-surfaces it is not even always true. For C^2-surfaces there is a general theorem guaranteeing their existence, but we do not have to invoke this theorem, since in the case of minimal surfaces we are able to give an elementary proof.

LEMMA 4.4. *Let S be a minimal surface. Every regular point of S has a neighborhood in which there exists a reparametrization of S in terms of isothermal parameters.*

Proof: By Lemma 1.2 we may first of all find a neighborhood of the regular point in which S may be represented in a non-parametric form. We then have equations (3.16) satisfied in some disk $(x_1 - a_1)^2 + (x_2 - a_2)^2 < R^2$. These equations imply the existence of functions $F(x_1, x_2)$, $G(x_1, x_2)$ in this disk, satisfying

$$(4.12) \qquad \frac{\partial F}{\partial x_1} = \frac{1 + |p|^2}{W} \ , \quad \frac{\partial F}{\partial x_2} = \frac{p \cdot q}{W} \ ;$$

$$\frac{\partial G}{\partial x_1} = \frac{p \cdot q}{W} \ , \quad \frac{\partial G}{\partial x_2} = \frac{1 + |q|^2}{W} \ .$$

If we set

$$(4.13) \qquad \xi_1 = x_1 + F(x_1, x_2), \quad \xi_2 = x_2 + G(x_1, x_2),$$

we find

$$\frac{\partial \xi_1}{\partial x_1} = 1 + \frac{1 + |p|^2}{W} \qquad \frac{\partial \xi_1}{\partial x_2} = \frac{p \cdot q}{W}$$

$$\frac{\partial \xi_2}{\partial x_1} = \frac{p \cdot q}{W} \qquad \frac{\partial \xi_2}{\partial x_2} = 1 + \frac{1 + |q|^2}{W} \ ,$$

and

$$J = \frac{\partial(\xi_1, \xi_2)}{\partial(x_1, x_2)} = 2 + \frac{2 + |p|^2 + |q|^2}{W} > 0 \ .$$

Thus the transformation (4.13) has a local inverse $(\xi_1, \xi_2) \to (x_1, x_2)$, and setting $x_k = f_k(x_1, x_2)$ for $k = 3, \dots, n$, we may represent the surface in terms of the parameters ξ_1, ξ_2. We find

$$\frac{\partial x_1}{\partial \xi_1} = \frac{W + 1 + |q|^2}{JW} , \qquad \frac{\partial x_2}{\partial \xi_1} = - \frac{p \cdot q}{JW} ,$$

$$\frac{\partial x_k}{\partial \xi_1} = \frac{W + 1 + |q|^2}{JW} p_k - \frac{p \cdot q}{JW} q_k , \qquad k = 3, \dots, n;$$

$$\frac{\partial x_1}{\partial \xi_2} = - \frac{p \cdot q}{JW} , \qquad \frac{\partial x_2}{\partial \xi_2} = \frac{W + 1 + |p|^2}{JW} ,$$

$$\frac{\partial x_k}{\partial \xi_2} = \frac{W + 1 + |p|^2}{JW} q_k - \frac{p \cdot q}{W} p_k , \quad k = 3, \dots, n.$$

It follows that with respect to the parameters ξ_1, ξ_2, we have

$$g_{11} = g_{22} = \left| \frac{\partial x}{\partial \xi_1} \right|^2 = \left| \frac{\partial x}{\partial \xi_2} \right|^2 = \frac{W}{J} = \frac{W^2}{2W + 2 + |p|^2 + |q|^2} \ ;$$

(4.14)

$$g_{12} = \frac{\partial x}{\partial \xi_1} \cdot \frac{\partial x}{\partial \xi_2} = 0 \ ,$$

so that ξ_1, ξ_2 are isothermal coordinates. ◆

COROLLARY. *Let* $x_k = f_k(x_1, x_2)$, $k = 3, \dots, n$, *define a minimal surface in non-parametric form. Then the* f_k *are real analytic functions of* x_1, x_2.

Proof: In a neighborhood of each point we can introduce the map (4.13) which gives local isothermal parameters ξ_1, ξ_2 for the surface. By Lemma 4.2, x_1 and x_2 are harmonic, hence real-analytic functions of ξ_1, ξ_2. Thus the inverse map $x_1, x_2 \to \xi_1, \xi_2$ is also real analytic. But each x_k is harmonic in ξ_1, ξ_2, hence a real analytic function of x_1, x_2.

Let us note in particular that every solution $f(x_1, x_2) \in C^2$ of the system (2.8) is automatically analytic. In the case $n = 3$ an elementary argument of this type was given by Müntz [1] and Radó [1], who considered, instead of (4.13), the mapping $\xi_1 = x_1$, $\xi_2 = G(x_1, x_2)$. This is a somewhat simpler mapping which also gives isothermal coordinates. However, the mapping (4.13) which was introduced in the case $n = 3$ by Nitsche [1] has additional properties which make it particularly useful, as we shall see in the following section.

We conclude this section with the following elementary lemma.

LEMMA 4.5. *Let a surface S be defined by $x(u)$, where u_1, u_2 are isothermal parameters, and let $\tilde{S}$ be a reparametrization of S defined by a diffeomorphism $u(\tilde{u})$. Then $\tilde{u}_1, \tilde{u}_2$ are also isothermal parameters if and only if the map $u(\tilde{u})$ is either conformal or anti-conformal.*

Proof: Since u_1, u_2 are isothermal, we have $g_{ij} = \lambda^2 \delta_{ij}$, and by (1.8), $\tilde{G} = \lambda^2 U^{\perp}U$. Thus $\tilde{u}_1, \tilde{u}_2$ isothermal $\Longleftrightarrow \tilde{g}_{ij} = \tilde{\lambda}^2 \delta_{ij} \Longleftrightarrow (\tilde{\lambda}/\lambda)U$ an orthogonal matrix $\Longleftrightarrow u(\tilde{u})$ is either conformal or anti-conformal. ♦

§5. *Bernstein's Theorem.*

In this section we shall prove several results related to Bernstein's Theorem. Although this theorem is of a global, rather than a local nature, we include it here before our general discussion of surfaces in the large, for two reasons. First, because the proof requires a purely elementary argument, and second, because Bernstein's Theorem provides the motivation for a number of the results that we discuss later on.

We begin with some elementary lemmas.

LEMMA 5.1. *Let* $E(x_1, x_2) \, \epsilon \, C^2$ *in a convex domain* D, *and suppose that the Hessian matrix*

$$\left(\frac{\partial^2 E}{\partial x_i \partial x_j} \right)$$

is positive definite. Define a mapping

(5.1) $$(x_1, x_2) \to (u_1, u_2), \text{ where } u_i = \frac{\partial E}{\partial x_i} \quad .$$

Then if x *and* y *are two distinct points of* D, *and if* u *and* v *are their respective image points under the map* (5.1), *the vectors* $y - x$ *and* $v - u$ *satisfy the equation*

(5.2) $$(v - u) \cdot (y - x) > 0 \, .$$

Proof: Let $G(t) = E(ty + (1-t)x)$, $0 \le t \le 1$. Then

$$G'(t) = \sum_{i=1}^{2} \left[\frac{\partial E}{\partial x_i} (ty + (1-t)x) \right] (y_i - x_i) \, ,$$

and

$$G''(t) = \sum_{i,j=1}^{2} \left[\frac{\partial^2 E}{\partial x_i \partial x_j} (ty + (1-t)x) \right] (y_i - x_i)(y_j - x_j)$$

$$> 0 \text{ for } 0 \le t \le 1 .$$

Hence $G'(1) > G'(0)$, or $\Sigma v_i(y_i - x_i) > \Sigma u_i(y_i - x_i)$, which is (5.2). ♦

LEMMA 5.2. (Lewy [1]). *Under the hypotheses of Lemma 5.1, if we define the map*

(5.3) $(x_1, x_2) \to (\xi_1, \xi_2)$, *by* $\xi_i(x_1, x_2) = x_i + u_i(x_1, x_2)$,

where $u_i(x_1, x_2)$ *is defined by* (5.1), *then for any two distinct points* x *and* y *of* D, *their images* ξ *and* η *satisfy*

(5.4) $(\eta - \xi) \cdot (y - x) > |y - x|^2 .$

Proof: Since $\eta - \xi = (y - x) + (v - u)$, this follows immediately from (5.2). ♦

COROLLARY. *Under the same hypotheses, we have*

(5.5) $|\eta - \xi| > |y - x| .$

Proof: By the Cauchy-Schwarz inequality,

$$|(\eta - \xi) \cdot (y - x)| \le |\eta - \xi| \, |y - x| ,$$

which applied to (5.4) yields (5.5).   ♦

LEMMA 5.3. *In the notation of the previous lemmas, if* D *is the disk* $x_1^2 + x_2^2 < R^2$, *then the map* (5.3) *is a diffeomorphism of* D *onto a domain* Δ *which includes a disk of radius* R *about* $\xi(0)$.

Proof: The map (5.3) is continuously differentiable, since $E(x_1, x_2) \in C^2$. If $x(t)$ is any differentiable curve in D, and $\xi(t)$ its image, then it follows from (5.5) that $|\xi'(t)| > |x'(t)|$, and hence the Jacobian is everywhere greater than 1. Thus the map is a local diffeomorphism, and by (5.5) it is one-to-one, hence a global diffeomorphism onto a domain Δ. We must show that Δ includes all points ξ such that $|\xi - \xi(0)| < R$. If Δ is the whole plane this is obvious. Otherwise there is a point ξ in the complement of Δ which minimizes the distance to $\xi(0)$. Let $\xi^{(k)}$ be a sequence of points in Δ which tend to ξ, and let $x^{(k)}$ be the corresponding points in D. The $x^{(k)}$ cannot have any point of accumulation in D since the image of such a point would then be the point ξ, contrary to the assumption that ξ is not in Δ. Thus $|x^{(k)}| \to R$ as $k \to \infty$, and since $|\xi^{(k)} - \xi(0)| > |x^{(k)}|$ by (5.5), it follows that $|\xi - \xi(0)| \geq R$, which proves the lemma. ◆

LEMMA 5.4. *Let* $f(x_1, x_2)$ *be a solution of the minimal surface equation* (3.10) *for* $x_1^2 + x_2^2 < R^2$. *Then using the notation* (3.8), (4.12), *the map* $(x_1, x_2) \to (\xi_1, \xi_2)$ *defined by* (4.13) *is a diffeomorphism onto a domain* Δ *which includes a disk of radius R about the point* $\xi(0)$.

Proof: It follows from equations (4.12) that there exists a function $E(x_1, x_2)$ in $x_1^2 + x_2^2 < R^2$ satisfying

$$(5.6) \qquad \frac{\partial E}{\partial x_1} = F, \qquad \frac{\partial E}{\partial x_2} = G .$$

Then $E(x_1, x_2) \in C^2$, and

$$\frac{\partial^2 E}{\partial x_1^2} = \frac{1 + |p|^2}{W} > 0, \quad \det \frac{\partial^2 E}{\partial x_1 \partial x_2} = \frac{\partial(F, G)}{\partial(x_1, x_2)} \equiv 1$$

by (3.12), (3.13) and (4.12). Thus the function $E(x_1, x_2)$ has a positive-definite Hessian matrix and we may apply Lemmas 5.1–5.3 to it. But by (5.6), the map (4.13) is just the map (5.3) applied to this function. Thus Lemma 5.4 follows immediately from Lemma 5.3. ♦

LEMMA 5.5. *Let* $f(x_1, x_2) \epsilon C^1$ *in a domain* D, *where* f *is real-valued. Necessary and sufficient that the surface* $S: x_3 = f(x_1, x_2)$ *lie on a plane is that there exist a nonsingular linear transformation* $(u_1, u_2) \to (x_1, x_2)$ *such that* u_1, u_2 *are isothermal parameters on* S.

Proof: Suppose such parameters u_1, u_2 exist. Introducing the functions $\phi_k(\zeta)$ by (4.6), $k = 1, 2, 3$, we see that ϕ_1 and ϕ_2 are constant since x_1 and x_2 are linear functions of u_1, u_2. But by (4.9), ϕ_3 must also be constant. This means that x_3 has constant gradient with respect to u_1, u_2, hence also with respect to x_1, x_2. Thus $f(x_1, x_2) = Ax_1 + Bx_2 + C$. Conversely, if f is of this form, it is easy to write down an explicit linear transformation yielding isothermal coordinates; for example, $x_1 = \lambda A u_1 + B u_2$, $x_2 = \lambda B u_1 - A u_2$, where $\lambda^2 = 1/(1 + A^2 + B^2)$. ♦

THEOREM 5.1. (Osserman [7]). *Let* $f(x_1, x_2)$ *be a solution of the minimal surface equation* (2.8) *in the whole* x_1, x_2-*plane. Then there exists a nonsingular linear transformation*

$$(5.7) \qquad \begin{aligned} x_1 &= u_1 \\ x_2 &= au_1 + bu_2, \qquad b > 0, \end{aligned}$$

such that (u_1, u_2) *are (global) isothermal parameters for the surface* S *defined by*

$$x_k = f_k(x_1, x_2), \qquad k = 3, \ldots, n .$$

COROLLARY 1. (Bernstein [3]). *In the case $n = 3$, the only solution of the minimal surface equation in the whole x_1, x_2-plane is the trivial solution, f a linear function of x_1, x_2.*

COROLLARY 2. *A bounded solution of equation (2.8) in the whole plane must be constant (for arbitrary n).*

COROLLARY 3. *Let $f(x_1, x_2)$ be a solution of (2.8) in the whole x_1, x_2-plane, and let $\tilde{S}$ be the surface defined by*

$$(5.8) \qquad x_k = \tilde{f}_k(u_1, u_2), \qquad k = 3, \dots, n$$

obtained by referring the surface S to the isothermal parameters given by (5.7). Then the functions

$$(5.9) \qquad \tilde{\phi}_k = \frac{\partial \tilde{f}_k}{\partial u_1} - i \frac{\partial \tilde{f}_k}{\partial u_2}, \qquad k = 3, \dots, n$$

are analytic functions of $u_1 + iu_2$ in the whole u_1, u_2-plane and satisfy

$$(5.10) \qquad \sum_{k=3}^{n} \tilde{\phi}_k^2 \equiv -1 - c^2, \qquad c = a - ib.$$

Conversely, *given any complex constant $c = a - ib$ with $b > 0$, and given any entire functions $\tilde{\phi}_3, \dots, \tilde{\phi}_n$ of $u_1 + iu_2$ satisfying (5.10), equations (5.9) may be used to define harmonic functions $\tilde{f}_k(u_1, u_2)$ and substituting u_1, u_2 as functions of x_1, x_2 from (5.7) into equations (5.8) yields a solution of the minimal surface equation (2.8) valid in the whole x_1, x_2-plane.*

Proof of Corollary 1: This is an immediate consequence of Theorem 5.1 and Lemma 5.5. ♦

Proof of Corollary 2. By Lemma 4.2, each x_k, for $k = 3, \ldots, n$, will be a bounded harmonic function of (u_1, u_2) in the whole u_1, u_2-plane, hence constant. ◆

Proof of Corollary 3: This is an immediate consequence of Lemma 4.3, using the fact that in view of (5.7),

$$\tilde{\phi}_1 \equiv \frac{\partial x_1}{\partial u_1} - i \frac{\partial x_1}{\partial u_2} \equiv 1, \quad \tilde{\phi}_2 = \frac{\partial x_2}{\partial u_2} - i \frac{\partial x_2}{\partial u_2} \equiv a - ib . \quad ◆$$

Proof of the theorem: We introduce the map (4.13), which is now defined in the entire x_1, x_2-plane. It follows from Lemma 5.4 that this map is a diffeomorphism of the x_1, x_2-plane onto the entire ξ_1, ξ_2-plane. We know from (4.14) that (ξ_1, ξ_2) are isothermal parameters on the surface S defined by $x_k = f_k(x_1, x_2)$, $k = 3, \ldots, n$. By Lemma 4.3, the functions

$$\phi_k(\zeta) = \frac{\partial x_k}{\partial \xi_1} - i \frac{\partial x_k}{\partial \xi_2} , \qquad k = 1, \ldots, n$$

are analytic functions of ζ. We note the identity

$$\text{Im}\{\bar{\phi}_1 \phi_2\} = - \frac{\partial(x_1, x_2)}{\partial(\xi_1, \xi_2)}$$

and since the Jacobian on the right is always positive, we deduce first that $\phi_1 \neq 0$, $\phi_2 \neq 0$ everywhere, and further that

$$\text{Im}\left\{\frac{\phi_2}{\phi_1}\right\} = \frac{1}{|\phi_1|^2} \text{Im}\{\bar{\phi}_1 \phi_2\} < 0 .$$

Thus the function ϕ_2/ϕ_1 is analytic in the whole ζ-plane, has negative imaginary part, and must therefore be a constant:

(5.11) $$\phi_2 = c\phi_1 ; \quad c = a - ib, \quad b > 0.$$

Taking the real and imaginary parts of equation (5.11), we find

(5.12)
$$\frac{\partial x_2}{\partial \xi_1} = a \frac{\partial x_1}{\partial \xi_1} - b \frac{\partial x_1}{\partial \xi_2}$$

$$\frac{\partial x_2}{\partial \xi_2} = b \frac{\partial x_1}{\partial \xi_1} + a \frac{\partial x_1}{\partial \xi_2} .$$

If we now introduce the transformation (5.7), equations (5.12) take the form

$$\frac{\partial u_1}{\partial \xi_1} = \frac{\partial u_2}{\partial \xi_2} , \quad \frac{\partial u_2}{\partial \xi_2} = - \frac{\partial u_1}{\partial \xi_2} ;$$

that is to say, the Cauchy-Riemann equations, expressing the condition that $u_1 + iu_2$ is a complex-analytic function of $\xi_1 + i\xi_2$. But this means, by Lemma 4.5, that (u_1, u_2) are also isothermal parameters, which proves the theorem. ◆

We may note that in the case $n = 3$ the introduction of the function $E(x_1, x_2)$ defined by (5.6) was suggested by E. Heinz (see Jörgens [1], p. 133), whereas the application of Lewy's map (5.3) to this function and the resulting elementary proof of Bernstein's Theorem is due to Nitsche [1, 2].

The point of Corollary 3 is that it provides a kind of representation theorem for all solutions of the minimal surface equation (2.8) over the whole x_1, x_2-plane. It is interesting to examine in detail the case $n = 4$. Then, as we have remarked at the end of section 2, one has in addition to the trivial solutions f_k linear, the solutions

(5.13) $$f_3 + if_4 = g(z), \quad z = x_1 + ix_2 ,$$

where g is an analytic function of z. For an arbitrary entire function $g(z)$, equation (5.13) defines a solution of the minimal surface equation in the whole x_1, x_2-plane. The same is true if one sets

$$(5.14) \qquad f_3 - if_4 = g(z) \, .$$

Now by Theorem 5.1, to every global solution $f_3(x_1, x_2), f_4(x_1, x_2)$, there corresponds a transformation (5.7) and entire functions $\tilde{\phi}_3(w)$, $\tilde{\phi}_4(w)$, where $w = u_1 + iu_2$, satisfying

$$(5.15) \qquad \tilde{\phi}_3^2 + \tilde{\phi}_4^2 \equiv -d, \qquad d = 1 + c^2 \, .$$

We consider two cases. First, if $c = \pm i$, then equation (5.10) reduces to

$$(5.16) \qquad \tilde{\phi}_3^2 + \tilde{\phi}_4^2 = 0; \qquad a = 0, \quad b = \pm 1 \, .$$

Thus the transformation (5.7) is either the identity transformation or a reflection: $w = z$, or $w = \bar{z}$, and equation (5.16) implies $\tilde{\phi}_4 = \pm i\tilde{\phi}_3$ which is equivalent to $f_3 + if_4$ an analytic function of z or $\bar{z}$. Thus the case $c = \pm i$ corresponds precisely to the special solutions (5.13) and (5.14).

In the second case, $c \neq \pm i$, we may write equation (5.15) in the form

$$(5.17) \qquad (\tilde{\phi}_3 + i\tilde{\phi}_4)(\tilde{\phi}_3 - i\tilde{\phi}_4) = -d, \qquad d \neq 0 \, .$$

Thus, each of the factors on the left is different from zero. In particular, the function $\tilde{\phi}_3 - i\tilde{\phi}_4$ is an entire function which never vanishes, and therefore is of the form

$$\tilde{\phi}_3 - i\tilde{\phi}_4 = e^{H(w)}$$

for some entire function $H(w)$. By (5.17) we have

$$\tilde{\phi}_3 + i\tilde{\phi}_4 = -de^{-H(w)},$$

and combining these two equations yields

(5.18) $\quad \tilde{\phi}_3 = \frac{1}{2}(e^{H(w)} - de^{-H(w)})$, $\quad \tilde{\phi}_4 = \frac{i}{2}(e^{H(w)} + de^{-H(w)})$.

We can thus describe explicitly in the case $n = 4$ all solutions of the minimal surface equation which are valid in the whole x_1, x_2-plane. They are either of the special form (5.13) or (5.14), or else they may be obtained by making an arbitrary linear transformation of the form (5.7), and inserting the constant $d = 1 + (a - ib)^2$ in (5.18) together with an arbitrary entire function $H(w)$.

Let us give a simple illustration. We choose $a = 0$, $b = 2$ in (5.7), and $H(w) = w$. Then

$$x_1 = u_1, \ x_2 = 2u_2, \ \tilde{\phi}_3 = \frac{1}{2}(e^w - 3e^{-w}), \ \tilde{\phi}_4 = \frac{i}{2}(e^w + 3e^{-w}),$$

$$x_3 = \text{Re} \int \tilde{\phi}_3 \, dw = \tfrac{1}{2}\cos u_2(e^{u_1} - 3e^{-u_1})$$

$$x_4 = \text{Re} \int \tilde{\phi}_4 \, dw = -\tfrac{1}{2}\sin u_2(e^{u_1} - 3e^{-u_1}) .$$

We thus obtain the surface in non-parametric form:

(5.19) $\quad x_3 = \tfrac{1}{2}\cos \dfrac{x_2}{2}(e^{x_1} - 3e^{-x_1}), \quad x_4 = -\tfrac{1}{2}\sin \dfrac{x_2}{2}(e^{x_1} - 3e^{-x_1})$

which, as one may verify by a direct computation, provides a global solution of the minimal surface equation for $n = 4$. This surface will be a useful example in connection with our general discussion later on.* (See p. 124.)

---

*See Appendix 3, Section 2.

§6. *Parametric Surfaces: Global Theory*

In the previous section we were able to obtain global results because of the special circumstance that we had a global parametrization in terms of two of the coordinates x_1, x_2. In the general case we have a surface covered by neighborhoods in each of which a parametrization of the form considered in section 1 is given. In order to study the whole surface, we have to first give precise definitions. We begin by recalling some facts about differentiable manifolds.

DEFINITIONS. An *n-manifold* is a Hausdorff space, each point of which has a neighborhood homeomorphic to a domain in E^n.

An *atlas* A for an n-manifold M is a collection of triples $(R_\alpha, O_\alpha, F_\alpha)$, where R_α is a domain in E^n, O_α is an open set on M, F_α is a homeomorphism of R_α onto O_α, and the union of all the O_α equals M. Each triple is called a *map*.

A manifold M is *orientable* if it possesses an atlas for which each transformation $F_\alpha^{-1} \circ F_\beta$ preserves orientation wherever it is defined. An *orientation* of M is the choice of such an atlas.

A C^r-*structure* on M is an atlas for which $F_\alpha^{-1} \circ F_\beta \in C^r$ wherever it is defined. A *conformal structure* on M is an atlas for which $F_\alpha^{-1} \circ F_\beta$ is a conformal map wherever it is defined.

REMARK. By "conformal," we mean "strictly conformal." Thus a conformal structure on M automatically provides an orientation of M.

Let M be an n-manifold with C^r-structure A, and $\tilde{M}$ an m-manifold with a C^r-structure $\tilde{A}$. A map $f: M \to \tilde{M}$ will be called a C^p-*map*, denoted $f \in C^p$, for $p \leq r$, if each map $\tilde{F}_\beta^{-1} \circ f \circ F_\alpha \in C^p$, wherever it is defined.

Let us note, in particular, that E^n has a canonical C^r-structure for all r, defined by letting A consist of the single triple $R_\alpha = O_\alpha = E^n$, F_α the identity map.

DEFINITION. A C^r-surface S in E^n is a 2-manifold M with a C^r-structure, together with a C^r-map $x(p)$ of M into E^n.

Let S be a C^r-surface in E^n, A the C^r-structure on the associated 2-manifold M, R_α a domain in the u-plane, and R_β a domain in the $\tilde{u}$-plane. Then the composition of F_α with the map $x(p)$ is a map $x(u)$ of R_α into E^n which defines a local surface in the sense of section 1. The corresponding map $x(\tilde{u})$ of R_β into E^n defines a local surface obtained from $x(u)$ by the change of parameters $u(\tilde{u}) = F_\alpha^{-1} \circ F_\beta$. Thus all local properties of surfaces which are independent of parameters are well defined on a global surface S given by the above definition. In particular, by a *point of S* we shall mean the pair $(p_0, x(p_0))$ where p_0 is a point of M, and we may speak of S being *regular* at a point, or of the *tangent plane* and the *mean curvature vector* of S at a point, etc.

The global properties of S will be defined simply to be those of M. Thus S will be called *orientable* if M is orientable, and an orientation of S is an orientation of M. Similarly for topological properties of S: S *compact, connected, simply connected*, etc.

We shall now make a convention which we shall adhere to throughout this paper. *All surfaces considered will be connected and orientable.* Obviously, if a surface is not connected one can consider separately each connected component, which is also a surface. As for non-orientable surfaces, they are certainly of interest, and in particular in the case of minimal surfaces, since they arise both analytically as elementary surfaces given by explicit formulas, such as Henneberg's surface (see for example Darboux [1], §226), and physically as soap-film surfaces of the type of the Möbius strip, bounded by a simple closed curve (see for example Courant [2], Chapter IV). However it is an elementary topological fact that to each non-orientable C^r-manifold M there corresponds an orientable C^r-manifold $\tilde{M}$ and a C^r-map $g: \tilde{M} \to M$, such that g is a local diffeomorphism, and the inverse image of each point of M consists of two points of $\tilde{M}$. Thus to each non-orientable surface $x(p): M \to E^n$ corresponds an orientable surface $x(\tilde{p}): \tilde{M} \to E^n$, where $x(\tilde{p}) = x(g(\tilde{p}))$, and many properties of the former can be read off immediately from corresponding properties of the latter. In particular, only such properties of non-orientable minimal surfaces will be derived in this survey.

DEFINITION. A regular C^2-surface S in E^n is a *minimal surface* if its mean curvature vector vanishes at each point.

LEMMA 6.1. *Let S be a regular minimal surface in E^n defined by a map $x(p): M \to E^n$. Then S induces a conformal structure on M.*

Proof: We are assuming by our convention that S is orientable. Let A be an oriented atlas of M. Let $\tilde{A}$ be the collection of all triples $(\tilde{R}_\alpha, \tilde{O}_\alpha, \tilde{F}_\alpha)$ such that $\tilde{R}_\alpha$ is a plane domain, $\tilde{O}_\alpha$ is an open set on M, $\tilde{F}_\alpha$ is a homeomorphism of $\tilde{R}_\alpha$ onto $\tilde{O}_\alpha$, $\tilde{F}_\beta^{-1} \circ \tilde{F}_\alpha$ preserves orientation wherever defined, and $x \circ \tilde{F}_\alpha: \tilde{R}_\alpha \to E^n$ defines a local surface in isothermal parameters. By Lemma 4.4 the union of the $\tilde{O}_\alpha$ equals M, so that $\tilde{A}$ is an atlas for M, and by Lemma 4.5 each $\tilde{F}_\alpha^{-1} \circ \tilde{F}_\beta$ is conformal wherever defined, so that $\tilde{A}$ defines a conformal structure on M. ♦

Let us note that the introduction of a conformal structure in this way may be carried out in great generality, since low-order differentiability conditions on S guarantee the existence of local isothermal parameters; however, the proof of their existence is far more difficult in the general case.

We now discuss some basic notions connected with conformal structure. We note first that if M has a conformal structure, then we can define all concepts which are invariant under conformal mapping. In particular, we can speak of *harmonic* and *subharmonic* functions on M, and (complex) analytic maps of one such manifold M into another $\tilde{M}$. A *meromorphic function* on M is a complex-analytic map of M into the Riemann sphere. The latter may be defined as the unit sphere: $|x| = 1$ in E^3, with the conformal structure defined by a pair of maps

$$(6.1) \quad F_1: \; x = \left(\frac{2u_1}{|w|^2 + 1}, \; \frac{2u_2}{|w|^2 + 1}, \; \frac{|w|^2 - 1}{|w|^2 + 1} \right), \qquad w = u_1 + iu_2$$

and

$$(6.2) \quad F_2: \; x = \left(\frac{2\tilde{u}_1}{|\tilde{w}|^2 + 1}, \; \frac{-2\tilde{u}_2}{|\tilde{w}|^2 + 1}, \; \frac{1 - |\tilde{w}|^2}{|\tilde{w}|^2 + 1} \right), \qquad \tilde{w} = \tilde{u}_1 + i\tilde{u}_2$$

The map F_1 is called *stereographic projection* from the point $(0,0,1)$, the image being the whole sphere minus this point. The map F_1^{-1} is given explicitly by

$$(6.3) \qquad\qquad F_1^{-1}: \; w = \frac{x_1 + ix_2}{1 - x_3}$$

and $F_1^{-1} \circ F_2$ is simply $w = 1/\tilde{w}$, a conformal map of $0 < |\tilde{w}| < \infty$ onto $0 < |w| < \infty$.

DEFINITION. A *generalized minimal surface* S in E^n is a non-constant map $x(p): M \to E^n$, where M is a 2-manifold with a conformal structure defined by an atlas $A = \{(R_\alpha, O_\alpha, F_\alpha)\}$, such that each coordinate function $x_k(p)$ is harmonic on M, and furthermore

(6.4)
$$\sum_{k=1}^{n} \phi_k^2(\zeta) \equiv 0 ,$$

where we set for an arbitrary α,

$$h_k(\zeta) = x_k(F_\alpha(\zeta)), \quad \phi_k(\zeta) = \frac{\partial h_k}{\partial \xi_1} - i \frac{\partial h_k}{\partial \xi_2} , \quad \zeta = \xi_1 + i\xi_2 .$$

Let us make the following comments concerning this definition. First of all, if S is a regular minimal surface, then using the conformal structure defined in Lemma 6.1, it follows from Lemma 4.3 that S is also a generalized minimal surface. Thus the theory of generalized minimal surfaces includes that of regular minimal surfaces. On the other hand, if S is a generalized minimal surface, then since the map $x(p)$ is non-constant, at least one of the functions $x_k(p)$ is non-constant, which implies that the corresponding analytic function $\phi_k(\zeta)$ can have at most isolated zeroes. Thus the equation

(6.5)
$$\sum_{k=1}^{n} |\phi_k^2(\zeta)| = 0$$

can hold at most at isolated points. Then again by Lemma 4.3, if we delete these isolated points from S, the remainder of the surface is a regular minimal surface. The points where equation (6.5) holds are called *branch points* of the surface. If we allow the case $n = 2$ in the definition of a generalized surface, we find that either $x_1 + ix_2$ or $x_1 - ix_2$ is a non-constant analytic function $f(\zeta)$. The points at which (6.5) holds are simply those where $f'(\zeta) = 0$, corresponding to branch points, in the classical sense, of the inverse mapping. In the case of arbitrary n, the difference between regular and generalized minimal surfaces consists in allowing the possibility of isolated branch points. There are both positive and negative aspects to enlarging the class of surfaces to be studied in this way. On the one hand, there are certain theorems one would like to prove for regular minimal surfaces, but which have, up to now, been settled only for generalized minimal surfaces.* The classical Plateau problem is a prime example. On the other hand, there are many theorems where the possible existence of branch points has no effect, and one may as well prove them for the wider class of generalized minimal surfaces. Let us give an example.

LEMMA 6.2. *A generalized minimal surface cannot be compact.*

Proof: Let S be a generalized minimal surface defined by a map $x(p): M \to E^n$. Then each coordinate function $x_k(p)$ is harmonic on M, and if M were compact $x_k(p)$ would attain its maximum, hence it would be constant, contradicting the assumption that the map $x(p)$ is non-constant. ◆

Finally, concerning the study of generalized minimal surfaces, let us note that precisely properties of the branch points themselves may be an object of investigation. See, for example, Bers [2] and Chen [1].

For the sake of brevity we make the following convention. We shall suppress the adjectives ''generalized'' and ''regular,'' and

---

*See Appendix 3, Section 1.

we shall refer simply to "minimal surfaces" except in those cases where either the statement would not be true without suitably qualifying it, or else where we wish to emphasize the fact that the surfaces in question are "regular" or "generalized."

We next give a brief discussion of Riemannian manifolds.

DEFINITION. Let M be an n-manifold with a C^r-structure defined by an atlas $A = \{(R_\alpha, O_\alpha, F_\alpha)\}$. A *Riemannian structure on M*, or a *C^q-Riemannian metric* is a collection of matrices G_α, where the elements of the matrix G_α are C^q-functions on O_α, $0 \leq q \leq r - 1$, and at each point the matrix G_α is positive definite, while for any α, β such that the map $u(\tilde{u}) = F_\alpha^{-1} \circ F_\beta$ is defined, the relation

$$(6.6) \qquad G_\beta = U^\mathsf{T} G_\alpha U$$

must hold, where U is the Jacobian matrix of the transformation $F_\alpha^{-1} \circ F_\beta$.

A *differentiable curve* on M is a differentiable map $p(t)$ of an interval $[a, b]$ of the real line into M.

The *length* of the curve $p(t)$, $a \leq t \leq b$, with respect to a given Riemannian metric is defined to be

$$(6.7) \qquad \int_a^b h(t)\,dt \,,$$

where for each t_0, $a \leq t_0 \leq b$, we choose an O_α such that $p(t_0) \in O_\alpha$, and we set

$$(6.8) \qquad h(t) = \left(\sum_{i,j=1}^n g_{ij}(p(t)) u_i'(t) u_j'(t) \right)^{1/2} , \quad G_\alpha = (g_{ij}) \,,$$

for t sufficiently near t_0, where u_1, u_2 are coordinates in R_α. By (6.6), the definition of $h(t)$ is independent of the choice of O_α.

A *divergent path* on M is a continuous map $p(t)$, $t \geq 0$, of the non-negative reals into M, such that for every compact subset Q of M, there exists t_0 such that $p(t) \notin Q$ for $t > t_0$.

If a divergent path is differentiable, we define its length to be

$$(6.9) \qquad\qquad \int_0^\infty h(t)\, dt$$

where $h(t)$ is again defined by (6.8).

DEFINITION. A manifold M is *complete* with respect to a given Riemannian metric if the integral (6.9) diverges for every differentiable divergent path on M.

The first investigation of complete Riemannian manifolds was made by Hopf and Rinow [1] in 1931. Since that time this subject has been studied extensively, and it is generally accepted that the notion of completeness is the most useful one for the global study of manifolds with a Riemannian metric. One of our aims in this survey will be to discuss in detail the structure of complete mini-may surfaces. First let us make the following observations.

Let a C^r-surface S in E^n be defined by a map $x(p): M \to E^n$. Then this map induces a Riemannian structure on M, where for each α we set $x(u) = x(F_\alpha(u))$, and we define G_α to be the matrix whose elements are

$$(6.10) \qquad\qquad g_{ij} = \frac{\partial x}{\partial x_i} \cdot \frac{\partial x}{\partial u_j} \ .$$

Then equation (6.6) is a consequence of (1.8), and the matrix G_α will be positive definite at each point where S is regular. Thus to each regular surface S in E^n corresponds a Riemannian

2-manifold M. We say that S is complete if M is complete with respect to the Riemannian metric defined by (6.10).

If S is a *generalized* minimal surface, then there will be iso- lated points at which the matrix defined by (6.10) will not be posi- tive definite. However, the function $h(t)$ defined by (6.5) is still a non-negative function, independent of the choice of a, and we may still define S to be *complete* if the integral (6.9) diverges for every divergent path.

We conclude this section by recalling some basic facts from the theory of 2-manifolds.

First of all, each 2-manifold M has a *universal covering sur- face* which consists of a simply-connected 2-manifold $\hat{M}$ and a map $\pi : \hat{M} \to M$, with the property that each point of M has a neighbor- hood V such that the restriction of π to each component of $\pi^{-1}(V)$ is a homeomorphism onto V. In particular, the map π is a local homeomorphism, and it follows that any structure on $M : C^r$, con- formal, Riemannian, etc. induces a corresponding structure on $\hat{M}$. It is not hard to show that *M is complete with respect to a given Riemannian metric if and only if $\hat{M}$ is complete with respect to the induced Riemannian metric.*

Suppose now that S is a minimal surface defined by a map $x(p) : M \to E^n$. We then have an associated simply-connected mini- mal surface $\hat{S}$, called the *universal covering surface* of S, defined by the composed map $x(\pi(\hat{p})) : \hat{M} \to E^n$. It follows that $\hat{S}$ is regu- lar, if and only if S is regular, and $\hat{S}$ is complete if and only if S is complete. Thus, many questions concerning minimal surfaces may be settled by considering only simply-connected minimal sur- faces. In that case we have the following important simplification.

LEMMA 6.3. *Every simply-connected minimal surface S has a reparametrization in the form* $x(\zeta): D \to E^n$, *where D is either the unit disk,* $|\zeta| < 1$, *or the entire ζ-plane.*

Proof: Let S be defined by $x(p): M \to E^n$. By Lemma 6.2, M is not compact, and by the Koebe uniformization theorem (see, for example, Ahlfors and Sario [1], III, 11, G) M is conformally equivalent to either the unit disk or the plane. The composed map $D \to M \to E^n$ gives the result. ♦

Finally, let us recall some terminology which distinguishes the two cases in Lemma 6.3.

A 2-manifold M with a conformal structure is called *hyperbolic* if there exists a non-constant negative subharmonic function on M, and *parabolic* otherwise. The function $\text{Re}\{\zeta - 1\}$ shows that the unit disk $|\zeta| < 1$ is hyperbolic, and it is not hard to show that the entire plan is parabolic.

There are many interrelations between the conformal structure, the topological structure, and the Riemannian structure of a surface. Furthermore, in the case of minimal surfaces, we shall see that each of these structures plays a role in studying the geometry of the surface.

§7. *Minimal Surfaces with Boundary.*

In this section we shall investigate some properties of minimal surfaces with boundary. We do not go into detail on Plateau's problem for the reasons mentioned in the introduction, but we include a brief discussion in the context of the present paper.

DEFINITION. A sequence of points p_k on a manifold M is *divergent* if it has no points of accumulation on M.

If S is a minimal surface defined by a map $x(p): M \to E^n$, the *boundary values* of S are the set of points of the form $\lim x(p_k)$, for all divergent sequences p_k on M.

REMARK. If M is a bounded domain in the plane, then a sequence p_k in M is divergent if and only if it tends to the boundary. If $x(p)$ extends to a continuous map of the closure $\overline{M}$, then the boundary values of S are the image of the boundary of M.

LEMMA 7.1. *Every minimal surface lies in the convex hull of its boundary values.*

Proof: Let S be a minimal surface defined by $x(p): M \to E^n$. Suppose that the boundary values of S lie in a half space

$$L(x) = \sum_{k=1}^{n} a_k x_k - b \leq 0.$$

The function $h(p) = L(x(p))$ is harmonic on M, and by the maximum principle, $h(p) \leq 0$ on M. Namely, if $\sup h(p) = m$, we may choose points p_k such that $h(p_k) \to m$. If the p_k have a point of accumulation in M, then $h(p)$ would assume its maximum at this point, hence be constant. But choosing an arbitrary divergent sequence q_k we would have $h(q_k) = m$ for all k, and $\overline{\lim} h(q_k) \leq 0$,

hence $m \leq 0$. On the other hand, if p_k is divergent, then again
$m = \lim h(p_k) \leq 0$. Thus $L(x(p)) \leq 0$ on M, so that S lies in the
half-space $L(x) \leq 0$. But the convex hull of the boundary values
is the intersection of all the half-spaces which contain them, and
S lies in this intersection.   ♦

LEMMA 7.2. *Let $x(u)$ be a minimal surface in isothermal parameters defined in a plane domain D. Then $x(u)$ cannot be constant along any line segment in D.*

Proof: Since the map $x(u)$ is one-to-one in the neighborhood
of a regular point, and since the branch points, if any, are isolated,
the result follows immediately.   ♦

LEMMA 7.3. (Reflection principle). *Let $x(u)$ be a minimal sur-
face in isothermal parameters defined in a semi-disk $D: |u| < \varepsilon$,
$u_2 > 0$. Suppose there exists a line L in space such that $x(u) \to L$
when $u_2 \to 0$. Then $x(u)$ can be extended to a generalized minimal
surface defined in the full disk $|u| < \varepsilon$. Furthermore this extended
surface is symmetric in L.*

Proof: By a rotation in E^n we may suppose that L is given by
the equations $x_k = 0$, $k = 1, \ldots, n-1$. Then the functions $x_k(u)$,
for $k = 1, \ldots, n-1$, may be extended by setting $x_k(u_1, 0) = 0$,
$x_k(u_1, u_2) = -x_k(u_1, -u_2)$. These extended functions are harmonic
in the full disk, by the reflection principle for harmonic functions.
Thus the functions

$$\phi_k = \frac{\partial x_k}{\partial u_1} - i \frac{\partial x_k}{\partial u_2} , \quad k = 1, \ldots, n-1$$

are analytic in the full disk and are pure imaginary on the real axis.

By the equation

$$\phi_n^2 = -\sum_{k=1}^{n-1} \phi_k^2 \quad,$$

we see that ϕ_n^2 extends continuously to the real axis and has non-negative real values there. It follows that ϕ_n extends continuously to the real axis and has real values there. By integration, x_n extends continuously to the real axis, and satisfies $\partial x_n / \partial u_2 = 0$ there. If we then set $x_n(u_1, u_2) = x_n(u_1, -u_2)$ in the lower half-disk we obtain the desired result. ◆

LEMMA 7.4. *Let* $x(u)$ *be a minimal surface in isothermal parameters defined in a disk* D. *Then* $x(u)$ *cannot tend to a single point along any boundary arc of* D.

Proof: If it did, we could apply Lemma 7.3, after a preliminary conformal map of D onto the upper half-plane, and extend the surface over a segment of the real axis. It would then be constant on this segment, contradicting Lemma 7.2. ◆

We now state the fundamental existence theorem concerning the existence of a minimal surface with prescribed boundary.[*]

THEOREM 7.1 (Douglas [1]). *Let* Γ *be an arbitrary Jordan curve in* E^n. *Then there exists a simply-connected generalized minimal surface bounded by* Γ.

We content ourselves here with an outline of the proof, based on important modifications due to Courant [1, 2]. We refer also to the versions given in the books of Radc [3], Lewy [2], and Garabedian [1].

First of all, let us give a precise statement of the conclusion.

---

[*]See Appendix 3, Section 1.

Let us use the following notation:

D is the unit disk: $u_1^2 + u_2^2 < 1$

$\bar{D}$ is the closure: $u_1^2 + u_2^2 \leq 1$

C is the boundary: $u_1^2 + u_2^2 = 1$.

The result is that there exists a map $x(u): \bar{D} \to E^n$, such that

  i) $x(u)$ is continuous on $\bar{D}$,

 ii) $x(u)$ restricted to D is a minimal surface,

iii) $x(u)$ restricted to C is a homeomorphism onto Γ.

This mapping is obtained by the method of minimizing the Dirichlet integral. To each map $x(u) \, \epsilon \, C^1$ in D, we denote by

$$\mathfrak{D}(x) = \iint_D \sum_{k=1}^{n} \left[\left(\frac{\partial x_k}{\partial u_1} \right)^2 + \left(\frac{\partial x_k}{\partial u_2} \right)^2 \right] du_1 \, du_2$$

its Dirichlet integral. In order to single out a suitable class of mappings in which the Dirichlet integral will attain a minimum, we consider *monotone* maps of C onto Γ; that is, maps such that if C is traversed once in the positive direction, then Γ is traversed once also in a given direction, although we allow arcs of C to map onto single points of Γ.

To a given Jordan curve Γ we associate the class H of maps $x(u): \bar{D} \to E^n$ having the following properties:

  a) $x(u)$ is continuous in $\bar{D}$;

  b) $x(u) \, \epsilon \, C^1$ in D ;

  c) $\mathfrak{D}(x) < \infty$;

  d) $x(u)$ restricted to C is a monotone map onto Γ.

It may well be that H is empty. However, we make the following assertions.

  1.   If Γ is piecewise differentiable (in particular, a polygon),

or if Γ is even rectifiable, then H is not empty.

2. Whenever H is not empty there exists a map $y(u) \, \epsilon \, H$ such that $\mathcal{D}(y) \leq \mathcal{D}(x)$ for all $x(u) \, \epsilon \, H$.

3. This map $y(u)$ satisfies the conditions i), ii), iii) above for a solution of our problem.

We now outline the proof of these assertions.

1. If we let s be the parameter of arc length on Γ, L its total length, and $t = 2\pi s/L$, then the map $x(e^{it}): C \to \Gamma$ can be extended by the Poisson integral applied to each coordinate $x_k(e^{it})$ to a map of $\overline{D} \to E^n$ which will be in H.

2. Let $d = \inf \mathcal{D}(x)$ for $x \, \epsilon \, H$, and let $x^\nu(u) \, \epsilon \, H$ such that $d_\nu = \mathcal{D}(x^\nu) \to d$. For each ν, define $\tilde{x}^\nu(u)$, by letting $\tilde{x}_k^\nu(u)$ be the harmonic function in D having the same boundary values as $x_k^\nu(u)$. Then $\tilde{x}^\nu(u) \, \epsilon \, H$, and hence $d \leq \mathcal{D}(\tilde{x}^\nu) \leq \mathcal{D}(x^\nu)$, by the property of harmonic functions to minimize the Dirichlet integral with given boundary values. Now fix three points on C and three points on Γ, and let $\hat{H}$ be the subset of H for which $x(u)$ maps the former onto the latter in a given order. By a linear fractional transformation of D, we can obtain from each $\tilde{x}^\nu(u)$ an $\hat{x}^\nu(u) \, \epsilon \, \hat{H}$, and $\mathcal{D}(\hat{x}^\nu) = \mathcal{D}(\tilde{x}^\nu)$. One then proves the fundamental lemma that boundary maps $\hat{x}^\nu: C \to \Gamma$ must be equicontinuous. We can therefore find a subsequence of the $\hat{x}^\nu$ which converge uniformly on C, hence in the closed disk $\overline{D}$, and the limit will be a map $y(u) \, \epsilon \, \hat{H}$, with $\mathcal{D}(y) = d$.

3. The map $y(u)$, being the uniform limit of harmonic functions in D is also harmonic in D. One shows by a variational argument that such a map must define a generalized minimal surface in D. Since the boundary correspondence is monotone, the only way it

could fail to be one-to-one is if a whole boundary arc maps onto a point, but by Lemma 7.4 this is impossible. Thus the map $y(u)$ satisfies conditions i), ii), iii) and solves the problem.

For the final step in the proof of Theorem 7.1, Douglas shows that if Γ is an arbitrary Jordan curve, it can be approximated by a sequence of polygons, and the corresponding sequence of minimal surfaces will tend to a surface which again satisfies i), ii), iii). ♦

REMARKS. 1. The restriction that the surface S be simply-connected is neither necessary, nor from a certain point of view, natural. In fact, if one preceeds to construct minimal surfaces physically, using a piece of wire and soap solution, one finds in many cases surfaces of higher genus and connectivity, including, among the simplest, the Möbius strip. For further discussion of this theory we refer to the fundamental paper of Douglas [2], and the book of Courant [2]. We limit ourselves here to mentioning two of the most striking facts concerning the order of complications that may arise. First of all there exist rectifiable Jordan curves Γ which bound a *non-denumerable* number of distinct minimal surfaces (Lévy [1]); second, there exist rectifiable Jordan curves Γ for which a surface of minimum area bounded by Γ must have infinite connectivity (Fleming [1]).

2. In the past few years an entirely new approach to Plateau's problem has been instituted, in which one seeks a minimum of area among a very general class of objects, instead of restricting the competition to surfaces, and then shows that there is a minimizing object which is in fact a surface. In a pioneering work of Reifenberg [1], the class of objects considered are compact subsets having a given boundary in a certain sense, and the 2-dimen-

sional Hausdorff measure is minimized. Two major advantages of
this method is that it can be applied to higher dimensional sub-
varieties, in which case one minimizes the m-dimensional Hausdorff measure, and that in the 2-dimensional case one obtains not
only a generalized minimal surface, but in fact a regular one. The
latter follows from the combined results of Reifenberg [1, 2, 3], and
in particular, from Theorem 4, p. 70 of [1] together with [3]. Fleming [2], basing on the methods of Federer and Fleming [1], shows
the existence of a regular oriented minimal surface bounded by an
oriented rectifiable Jordan curve Γ. ´ See also the discussion in
Almgren [3].

3. One of the major open problems concerning Theorem 7.1 is
whether there, in fact, exists a *regular simply-connected* minimal
surface bounded by an arbitrary Jordan curve Γ.* This was actually the way Plateau's problem was originally envisaged, and in this
form it is still unsolved. In the following lemmas we give some information concerning cases in which the surface constructed in
Theorem 7.1 must, in fact, be regular.

LEMMA 7.5 (Rado [3], III.7). *Let* $h(u_1, u_2)$ *be non-constant
harmonic in the unit disk* D, *continuous in the closure* $\bar{D}$. *Suppose
that the gradient of* h *vanishes at a point* (a_1, a_2) *in* D. *Then if*
$h(a_1, a_2) = b$, *there are at least four distinct points on the boundary
of* D *where* h *takes on the value* b.

Proof: Let Δ be a component of the set of points in $\bar{D}$ where
$h > b$. Then at boundary points of Δ interior to D we have $h = b$.
Since $h \not\equiv b$, each component Δ must have boundary points on the
boundary of D with $h > b$. Similarly for each component of the set
where $h < b$. In the neighborhood of the point (a_1, a_2) there are at

---

*See Appendix 3, Section 1.

least two components of the set $h > b$, and two of the set $h < b$. If the former were part of a single component Δ in which $h > b$, then we could find an arc joining them which could be completed to a Jordan curve in D through (a_1, a_2), and this curve would completely enclose a domain in which $h < b$, contradicting our previous conclusion that such a domain must go to the boundary of D. Thus there are at least two distinct components where $h > b$ and two where $h < b$, and each of these intersects the boundary at points where $h \neq b$. But if there were at most three boundary points where $h = b$, there would be at most three complementary arcs on each of which $h > b$ or $h < b$, hence at most three components of sets $h > b$ and $h < b$, which contradicts what we have shown.   ♦

LEMMA 7.6. *Let Γ be a Jordan curve in E^n, and let $x(u)$ be a simply-connected generalized minimal surface bounded by Γ in the sense of Theorem 7.1. Then either $x(u)$ is in fact a regular minimal surface or else Γ has the property that for some point in E^n, every hyperplane through this point intersects Γ in at least four distinct points.*

Proof: We use the notation of the proof of Theorem 7.1. Thus $x(u)$ is continuous in $\overline{D}$, and defines a generalized minimal surface in D. Suppose that at some point (a_1, a_2) of D, $x(u)$ fails to be regular. Then at this point all the functions

$$\phi_k = \frac{\partial x_k}{\partial u_1} - i \frac{\partial x_k}{\partial u_2}$$

must vanish. If $L(x) \equiv \Sigma a_k x_k + c = 0$ is the equation of an arbitrary hyperplane through the point $x(a_1, a_2)$, then $h(u_1, u_2) = L(x(u_1, u_2))$ is a function satisfying the hypotheses of Lemma 7.5, with $b = 0$. It follows that there are at least four distinct points on Γ where $L(x) = 0$, which proves the lemma. ◆

This lemma shows that unless Γ is fairly complicated, the surface that it bounds has no branch points. In many specific cases we may assert the existence of regular simply-connected surfaces bounded by Γ, thus giving in those cases a complete solution of Plateau's problem.

THEOREM 7.2.[*] *Let D_1 be a bounded convex domain in the (x_1, x_2)-plane, and let C_1 be its boundary. Let $g_k(x_1, x_2)$, $k = 3, \ldots, n$, be arbitrary continuous functions on C_1. Then there exists a solution*

$$f(x_1, x_2) = (f_3(x_1, x_2), \ldots, f_n(x_1, x_2))$$

of the minimal surface equation (2.8) in D_1, such that $f_k(x_1, x_2)$ takes on the boundary values $g_k(x_1, x_2)$.

Proof: The boundary C_1 of D_1 is a Jordan curve(even rectifiable) and the functions $g_k(x_1, x_2)$ define a Jordan curve Γ in E^n which projects onto C_1. By Theorem 7.1, there exists a continuous map $x(u)\colon \overline{D} \to E^n$ of the unit disk $|u| \leq 1$ which defines a minimal surface in the interior D, and takes the boundary C homeomorphically onto Γ. It follows that the map $(u_1, u_2) \to (x_1, x_2)$ is a continuous map of $\overline{D}$ which is harmonic in the interior and maps C homeomorphically onto C_1. By Lemma 7.1, the image of D must lie in D_1. Furthermore, the Jacobian never vanishes in D. Namely, if at some point (a_1, a_2) in D the Jacobian were to vanish, then the rows of the Jacobian matrix would be linearly dependent at this point, i.e., for suitable constants λ_1, λ_2, we would have

---

[*]See Appendix 3, Section 5.

$$\lambda_1 \frac{\partial x_1}{\partial u_1} + \lambda_2 \frac{\partial x_2}{\partial u_1} = 0, \qquad \lambda_1 \frac{\partial x_1}{\partial u_2} + \lambda_2 \frac{\partial x_2}{\partial u_2} = 0 \, .$$

Then the function $h(u_1, u_2) = \lambda_1 x_1(u_1, u_2) + \lambda_2 x_2(u_1, u_2)$ would satisfy the hypotheses of Lemma 7.5 and it would follow that $\lambda_1 x_1 + \lambda_2 x_2$ would take on the same value at four distinct points of C_1, which is impossible by the convexity of D_1. We therefore conclude that the map $(u_1, u_2) \to (x_1, x_2)$ is a local diffeomorphism in D, and since it maps C homeomorphically onto C_1, it is a global diffeomorphism of D onto D_1. We may therefore express the x_k, $k = 3, \ldots, n$ as functions of x_1, x_2, and these functions will satisfy the conclusion of our theorem. ♦

Let us note in conclusion that except for the proof of Theorem 7.1, all the results in this section have been adapted from the treatment for the case $n = 3$ in the book of Rado [3].

§8. Parametric Surfaces in E^3. The Gauss Map.

In everything we have done up to now, there is either no difference at all, or else very little difference, between the classical
case of three dimensions and the case of arbitrary n. We shall
now discuss several results which either have not been extended
to arbitrary n, or else require a much more elaborate discussion to
do so.

We start with the important observation that for the case $n = 3$
we are able to describe explicitly all solutions of the equation

$$(8.1) \qquad \phi_1^2 + \phi_2^2 + \phi_3^2 = 0 .$$

LEMMA 8.1. *Let D be a domain in the complex ζ-plane, $g(\zeta)$
an arbitrary meromorphic function in D and $f(\zeta)$ an analytic function in D having the property that at each point where $g(\zeta)$ has a
pole of order m, $f(\zeta)$ has a zero of order at least $2m$. Then the
functions*

$$(8.2) \qquad \phi_1 = \tfrac{1}{2} f(1-g^2), \quad \phi_2 = \tfrac{i}{2} f(1+g^2), \quad \phi_3 = fg$$

*will be analytic in D and satisfy (8.1). Conversely, every triple
of analytic functions in D satisfying (8.1) may be represented in
the form (8.2), except for $\phi_1 \equiv i\phi_2$, $\phi_3 \equiv 0$.*

Proof: That the functions (8.2) satisfy (8.1) is a direct calculation. Conversely, given any solution of (8.1), we set

$$(8.3) \qquad f = \phi_1 - i\phi_2, \quad g = \frac{\phi_3}{\phi_1 - i\phi_2} .$$

If we write (8.1) in the form

$$(8.4) \qquad (\phi_1 - i\phi_2)(\phi_1 + i\phi_2) = -\phi_3^2 ,$$

we find

(8.5) $$\phi_1 + i\phi_2 = -\frac{\phi_3^2}{\phi_1 - i\phi_2} = -fg^2 .$$

Combining (8.3) and (8.5) yields (8.2). The condition relating the zeros of f and the poles of g must obviously hold, since otherwise by equation (8.5), $\phi_1 + i\phi_2$ would fail to be analytic. This representation can fail only if the denominator in the expression for g in (8.3) vanishes identically. In this case we have by (8.4) that $\phi_3 \equiv 0$, which is the exceptional case mentioned.

LEMMA 8.2. *Every simply-connected minimal surface in* E^3 *can be represented in the form*

(8.6) $$x_k(\zeta) = \text{Re}\left\{ \int_0^\zeta \phi_k(z)dz \right\} + c_k , \qquad k = 1, 2, 3$$

where the ϕ_k *are defined by* (8.2), *the functions* f *and* g *having the properties stated in Lemma 8.1, the domain* D *being either the unit disk or the entire plane, and the integral being taken along an arbitrary path from the origin to the point* ζ. *The surface will be regular if and only if* f *satisfies the further property that it vanishes only at the poles of* g, *and the order of its zero at such a point is exactly twice the order of the pole of* g.

Proof: By Lemma 6.3, the surface may be represented in the form $x(\zeta): D \to E^3$ where D is either the disk or the plane, the coordinates x_k being harmonic in ζ. If we set

$$\phi_k = \frac{\partial x_k}{\partial \xi_1} - i\frac{\partial x_k}{\partial \xi_2} , \qquad \zeta = \xi_1 + i\xi_2 ,$$

then these functions will be analytic and (8.6) will hold (the intetegral being independent of path). For a generalized minimal surface the equation (8.1) must hold and by Lemma 8.1 we have the representation (8.2). The surface will fail to be regular if and only if all the ϕ_k vanish simultaneously, which happens precisely when $f = 0$ where g is regular or when $fg^2 = 0$ where g has a pole. ◆

Let us note that representations of the form (8.2), (8.6) were first given by Enneper and Weierstrass, and have played a major role in the theory of minimal surfaces in E^3. For one thing, they allow us to construct a great variety of specific surfaces having interesting properties. For example, the most obvious choice: $f \equiv 1$, $g(\zeta) = \zeta$, leads to the surface known as *Enneper's surface*. More important, this representation allows us to obtain general theorems about minimal surfaces by translating the statements into corresponding statements about analytic functions. In order to do this we must first express the basic geometric quantities associated with the surface in terms of the functions f, g.

First of all, the tangent plane is generated by the vectors

$$\frac{\partial x}{\partial \xi_1}, \ \frac{\partial x}{\partial \xi_2}, \ \text{where} \ \frac{\partial x}{\partial \xi_1} - i \, \frac{\partial x}{\partial \xi_2} = (\phi_1, \phi_2, \phi_3).$$

It follows that

(8.7) $$g_{ij} = \lambda^2 \delta_{ij},$$

where

$$\lambda^2 = |\frac{\partial x}{\partial \xi_1}|^2 = |\frac{\partial x}{\partial \xi_2}|^2 = \frac{1}{2} \Sigma \ |\phi_k|^2 = \left[\frac{|f|(1+|g|^2)}{2} \right]^2 .$$

Furthermore

$$\frac{\partial x}{\partial \xi_1} \times \frac{\partial x}{\partial \xi_2} = \text{Im}\{(\phi_2 \bar{\phi}_3, \ \phi_3 \bar{\phi}_1, \ \phi_1 \bar{\phi}_2)\}$$

and substituting in (8.2) we arrive at the expression

$$\frac{\partial x}{\partial \xi_1} \times \frac{\partial x}{\partial \xi_2} = \frac{|f|^2(1+|g|^2)}{4}(2 \operatorname{Re}\{g\},\ 2 \operatorname{Im}\{g\},\ |g|^2 - 1)\ .$$

From this it follows that

$$\left|\frac{\partial x}{\partial \xi_1} \times \frac{\partial x}{\partial \xi_2}\right| = \left[\frac{|f|(1+|g|^2)}{2}\right]^2 = \lambda^2$$

and

$$(8.8) \qquad N = \frac{\dfrac{\partial x}{\partial \xi_1} \times \dfrac{\partial x}{\partial \xi_2}}{\left|\dfrac{\partial x}{\partial \xi_1} \times \dfrac{\partial x}{\partial \xi_2}\right|} = \left(\frac{2\operatorname{Re}\{g\}}{|g|^2+1},\ \frac{2\operatorname{Im}\{g\}}{|g|^2+1},\ \frac{|g|^2-1}{|g|^2+1}\right)$$

is the unit normal to the surface with the standard orientation.

Now given an arbitrary regular surface $x(u)$ in E^3, one defines the *Gauss map* to be the map

$$(8.9) \qquad x(u) \to N(u) = \frac{\dfrac{\partial x}{\partial u_1} \times \dfrac{\partial x}{\partial u_2}}{\left|\dfrac{\partial x}{\partial u_1} \times \dfrac{\partial x}{\partial u_2}\right|}$$

of the surface into the unit sphere.

LEMMA 8.3. *If* $x(\zeta): D \to E^3$ *defines a regular minimal surface in isothermal coordinates, then the Gauss map* $N(\zeta)$ *defines a complex analytic map of* D *into the unit sphere considered as the Riemann sphere.*

Proof: Formula (8.8) compared with formula (6.1) for stereographic projection shows that the Gauss map $x(\zeta) \to N(\zeta)$ followed by stereographic projection from the point $(0, 0, 1)$ yields the mero-morphic function $g(\zeta)$. ♦

Let us note that in general the Gauss map cannot be defined if a surface is not regular. However, for a generalized minimal surface the Gauss map extends continuously, and even analytically, to the branch points, the normal N being given by the right-hand side of (8.8).

LEMMA 8.4. *Let* $x(\zeta)$: $D \to E^3$ *define a generalized minimal surface* S, *where* D *is the entire* ζ-*plane. Then either* $x(\zeta)$ *lies on a plane, or else the normals to* S *take on all directions with at most two exceptions.*

Proof: To the surface S we associate the function $g(\zeta)$ which fails to be defined only if $\phi_1 \equiv i\phi_2$, $\phi_3 \equiv 0$. But in this case x_3 is constant and the surface lies in a plane. Otherwise $g(\zeta)$ is meromorphic in the entire ζ-plane, and by Picard's theorem it either takes on all values with at most two exceptions, or else is constant. But by (8.8) the same alternative applies to the normal N, and in the latter case S lies on a plane. ♦

LEMMA 8.5. *Let* $f(z)$ *be an analytic function in the unit disk* D *which has at most a finite number of zeros. Then there exists a divergent path* C *in* D *such that*

$$(8.10) \qquad \int_C |f(z)| \, |dz| < \infty \,.$$

Proof: Suppose first that $f(z) \neq 0$ in D. Define

$$w = F(z) = \int_0^z f(\zeta) d\zeta \,.$$

Then $F(z)$ maps $|z| < 1$ onto a Riemann surface which has no branch points. If we let $z = G(w)$ be that branch of the inverse function satisfying $G(0) = 0$, then since $|G(w)| < 1$, there is a largest disk $|w| < R < \infty$ in which $G(w)$ is defined. There must then be a point w_0 with $|w_0| = R$ such that $G(w)$ cannot be extended to a neighborhood of w_0. Let L be the line segment $w = tw_0$, $0 \le t < 1$, and let C be the image of L under $G(w)$. Then C must be a divergent path, since otherwise there would be a sequence $t_n \to 1$ such that the corresponding sequence of points z_n on C would converge to a point z_0 in D. But then $F(z_0) = w_0$, and since $F'(z_0) = f(z_0) \ne 0$, the function $G(w)$ would be extendable to a neighborhood of w_0. Thus the path C is divergent, and we have

$$\int_C |f(z)| \, |dz| = \int_0^1 |f(z)| \, \left| \frac{dz}{dt} \right| dt = \int_0^1 \left| \frac{dw}{dt} \right| dt = R < \infty \, .$$

This proves the lemma if $f(z)$ has no zeros. But if it has a finite number of zeros, say of order ν_k at the points z_k, then the function

$$f_1(z) = f(z) \, \Pi \left(\frac{1 - \bar{z}_k z}{z - z_k} \right)^{\nu_k}$$

never vanishes, and by the above argument there exists a divergent path C such that $\int_C |f_1(z)| \, |dz| < \infty$. But $|f(z)| < |f_1(z)|$ throughout D, and (8.10) follows. ◆

THEOREM 8.1. *Let S be a complete regular minimal surface in E^3. Then either S is a plane or else the normals to S are everywhere dense.*

Proof: Suppose that the normals to S are not everywhere dense. Then there exists an open set on the unit sphere which is not intersected by the image of S under the Gauss map. By a rotation in space we may assume that the point $(0, 0, 1)$ is in this open set. Then the unit normals $N = (N_1, N_2, N_3)$ satisfy $N_3 \leq \eta < 1$. The same is true of the universal covering surface $\hat{S}$ of S, which may be represented in the form $x(\zeta) : D \to E^3$ where D is the plane or the unit disk. But D cannot be the unit disk, because by (8.8), $N_3 \leq \eta < 1 \Longrightarrow |g(\zeta)| \leq M < \infty$, and since $\hat{S}$ is regular $f(\zeta)$ cannot vanish. But by (8.7) the length of any path C would be

$$\int_C \lambda |d\zeta| = \frac{1}{2} \int_C |f|(1 + |g|^2)|d\zeta| < \frac{1 + M^2}{2} \int_C |f| \, |d\zeta|$$

and by Lemma 8.5 there would exist a divergent path C for which this integral converges, and the surface would not be complete. Thus D is the entire plane, and since the normals omit more than two points, it follows from Lemma 8.4 that $\hat{S}$ must lie on a plane. The same is then true of S, and since it is complete, S must be the whole plane. ◆

Let us note that Theorem 8.1 has as an immediate consequence the theorem of Bernstein. In fact, a non-parametric minimal surface in E^3 defined over the whole x_1, x_2-plane is a complete regular surface whose normals are contained in a hemisphere, hence it must be a plane.

On the other hand, Theorem 8.1 leads naturally to the question, given a complete minimal surface S, not a plane, what can be said about the size of the set of points on the sphere omitted by the Gauss map of S? Theorem 8.1 tells us that, at least if S is regular, the omitted set cannot contain a neighborhood of any point. A much stronger result can be obtained by introducing the notion of sets of *logarithmic capacity zero.* We may define these to be closed sets on the sphere whose complement is parabolic in the sense of conformal structure. (See the discussion at the end of section 6.) It is well known that parabolicity is equivalent to the non-existence of a Green's function. (See, for example, the book of Ahlfors and Sario [1], IV 6 and IV 22.)

LEMMA 8.6. *Let D be a domain in the complex w-plane. The complement E of D on the Riemann sphere has logarithmic capacity zero if and only if the function $\log(1+|w|^2)$ has no harmonic majorant in D.*

Proof: Suppose first that there exists a harmonic function $h(w)$ in D such that $\log(1+|w|^2) \leq h(w)$ everywhere. Then $-h(w)$ is a negative harmonic function in D, and hence E has positive logarithmic capacity. Conversely, if E has positive logarithmic capacity, then for any point w_0 in D there exists a Green's function $G(w, w_0)$ with pole at w_0. By definition, $G(w, w_0) + \log|w-w_0| = h(w)$, where $h(w)$ is harmonic in D, and $G(w, w_0) > 0$, so that $\log|w-w_0| < h(w)$. The function $\log[(1+|w|^2)/|w-w_0|^2]$ is continuous on the compact set E, hence has a finite maximum M. Thus if w_1 is any boundary point of D, we have

$$\overline{\lim_{w \to w_1}} \, [\log(1+|w|^2) - 2h(w)] \leq \overline{\lim_{w \to w_1}} \, [\log(1+|w|^2) - 2\log|w-w_0|]$$

$$\leq M \, .$$

But $\log(1+|w|^2)$ is subharmonic in D, and by the maximum principle we have $\log(1+|w|^2) \leq 2h(w) + M$ throughout D. ◆

THEOREM 8.2.[*] *Let* S *be a complete regular minimal surface in* E^3. *Then either* S *is a plane, or else the set* E *omitted by the image of* S *under the Gauss map has capacity zero.*

Proof: If S is not a plane, then the image of S under the Gauss map is an open connected set on the sphere, and the complement of the image is therefore a compact set[*] E. If E is empty there is nothing to prove. Otherwise we may assume that the set E includes the point $(0, 0, 1)$, after a preliminary rotation of coordinates. Again we may pass to the universal covering surface $\hat{S}$ of S, whose Gauss map omits the same set E. The surface S is given by a map $x(\zeta): D \to E^3$, where D is either the plane or the unit disk. In the former case we know that the set E can contain at most two points and hence certainly has capacity zero. Let us examine the case where D is the disk $|\zeta| < 1$. We have the associated function $g(\zeta)$ which is analytic in D, and by the regularity of $\hat{S}$, $f(\zeta) \neq 0$ in D. Suppose now that E did not have capacity zero. Then in the image D_1 of D under the map $w = g(\zeta)$, there would be a harmonic function $h(w)$ majorizing $\log(1 + |w|^2)$. Then $h(g(\zeta))$ is harmonic in D, and is the real part of an analytic function $G(\zeta)$ in D. Finally, $F(\zeta) = e^{G(\zeta)}$ is analytic in D and never zero. For an arbitrary path C in D, we have the length

$$\int_C \lambda |d\zeta| = \frac{1}{2} \int_C |f|(1 + |g|^2)|d\zeta| \leq \frac{1}{2} \int_C |fF| \, |d\zeta| \; .$$

But the function $f(\zeta)F(\zeta)$ never vanishes in D, and by Lemma 8.5 there would be a divergent path C for which the integral on the right converges, and the surface S would not be complete. Thus the set E must in fact have capacity zero, and the theorem is proved. ◆

---

[*]See Appendix 3, Section 4.

THEOREM 8.3. *Let* E *be an arbitrary set of* k *points on the unit sphere, where* $k \leq 4$. *Then there exists a complete regular minimal surface in* E^3 *whose image under the Gauss map omits precisely the set* E.

Proof: By a rotation we may assume that the set E contains the point $(0,0,1)$. If this is only point, then Enneper's surface defined earlier (by setting $f(\zeta) = 1$, $g(\zeta) = \zeta$) solves the problem. Otherwise let the other points of E correspond to the points w_m, $m = 1, \ldots, k-1$, under stereographic projection. If we set

$$f(\zeta) = \frac{1}{\displaystyle\prod_{m=1}^{k-1} (\zeta - w_m)}, \quad g(\zeta) = \zeta,$$

and use the representation (8.2), (8.6) in the whole ζ-plane minus the points w_m, we obtain a minimal surface whose normals omit precisely the points of E, by (8.8), and which is complete, because a divergent path C must tend either to ∞ or to one of the points w_m, and in either case, we have

$$\int_C \lambda |d\zeta| = \frac{1}{2} \int_C |f|(1+|g|^2)|d\zeta| = \infty .$$

We may note that the integrals (8.6) may not be single-valued, but by passing to the universal covering surface we get a single-valued map defining a surface having the same properties.   ◆

Let us review briefly the historical development of the above theorems. Theorem 8.1, with the additional assumption that S be simply connected was conjectured by Nirenberg as a natural generalization of Bernstein's theorem, and it was proved in Osserman[1]. The presentation given here follows that of Osserman [3], where it is observed that simple connectivity is irrelevant and where the formulas (8.2), (8.6), (8.8) are used to reduce geometric statements of this kind to purely analytic ones. Lemma 8.5 is given there for the case $f(z) \neq 0$, and Theorem 8.2 is stated as a conjecture. At the end of the paper there is given a proof of Theorem 8.2 due to Ahlfors, who observed that using the above-mentioned reduction to analytic functions the result followed from a theorem of Nevanlinna. Ahlfors also suggested the reasoning which allows one to include in Lemma 8.5 functions $f(z)$ with a finite number of zeros. This allows us to make the following geometric conclusion: *Theorems 8.1 and 8.2 remain valid for generalized minimal surfaces, provided that they are simply connected and have only a finite number of branch points.* Furthermore, Ahlfors observed that Lemma 8.5 continues to hold if $f(z)$ has an infinite number of zeros, provided that their Blaschke product is convergent. Thus Theorems 8.1 and 8.2 continue to hold for a certain class of generalized minimal surfaces which have an infinite number of branch points, but it is not clear how to characterize this class geometrically. On the other hand, let us note the following. *There exist complete generalized minimal surfaces, not lying in a plane, whose Gauss map lies in an arbitrarily small neighborhood on the sphere.* In fact, we need only choose D to be the unit disk $|\zeta| < 1$, $g(\zeta) = \varepsilon \zeta$ for any given $\varepsilon > 0$, and $f(\zeta)$ a function analytic in D such that $\int_C |f(\zeta)| \, |d\zeta| = \infty$ for every divergent path C. Such functions $f(\zeta)$ may be constructed in a variety of ways.

Returning to Theorem 8.2, the proof given here, which does not depend on Nevanlinna's theory of functions of bounded characteristic, is taken from Osserman [4]. Theorem 8.3 is due to Voss [1].

An example of a complete surface whose normals omit 4 directions had been given earlier in Osserman [3], and it was later observed in Osserman [6], that the classical minimal surface of Scherk provides still another example.

The obvious question which arises when comparing Theorems 8.2 and 8.3 is the exact size of the set E omitted by the normals.* Specifically the following:

Problems 1. Do there exist complete regular minimal surfaces whose image under the Gauss map covers all of the sphere except for any arbitrary finite set of points E given in advance?

2. Does there exist a complete regular minimal surface whose image under the Gauss map is the complement of an infinite set E of capacity zero?

---

*See Appendix 3, Section 4.

§9. Surfaces in E^3. Gauss Curvature and Total Curvature.

We continue our study of minimal surfaces in E^3, using the representation (8.2).

The second fundamental form was defined in (1.27) with respect to an arbitrary normal vector N. Since it is linear in N, and since in E^3 the normal space at each point is one-dimensional, it is sufficient to define $b_{ij}(N)$ for a single normal N which is usually chosen to be that in (8.8). By a calculation one finds the following expression for the second fundamental form for a minimal surface in the representation (8.2):

$$(9.1) \quad \Sigma \, b_{ij} \frac{d\xi_i}{dt} \frac{d\xi_j}{dt} = \text{Re}\left\{ -fg'\left(\frac{d\zeta}{dt}\right)^2 \right\}; \qquad \zeta = \xi_1 + i\xi_2 \ .$$

Since by (8.7) we have

$$\Sigma \, g_{ij} \frac{d\xi_i}{dt} \frac{d\xi_j}{dt} = \left[\frac{|f|(1+|g|^2)}{2}\right]^2 \left|\frac{d\zeta}{dt}\right|^2 \ ,$$

it follows by (1.28) that the normal curvature is given by

$$\left[\frac{2}{|f|(1+|g|^2)}\right]^2 \text{Re}\{-fg' \, e^{2i\alpha}\} \ ; \quad \frac{d\zeta}{dt} = \left|\frac{d\zeta}{dt}\right| e^{i\alpha} \ .$$

The maximum and minimum of this expression, as α varies from 0 to 2π, were defined in (1.30) to be the principal curvatures. They are obviously

$$(9.2) \quad k_1 = \frac{4|g'|}{|f|(1+|g|^2)^2} \ , \qquad k_2 = \frac{-4|g'|}{|f|(1+|g|^2)^2} \ .$$

Now for an arbitrary regular C^2-surface in E^3 one defines the *Gauss curvature* K at a point as the product of the principal curvatures:

(9.3) $$K = k_1 k_2 .$$

LEMMA 9.1. *Let* $x(\zeta): D \to E^3$ *define a minimal surface. Then using the representation* (8.2), *the Gauss curvature at each point is given by*

(9.4) $$K = - \left[\frac{4\,|g'|}{|f|(1+|g|^2)^2} \right]^2 .$$

Proof: (9.2) and (9.3). ◆

COROLLARY. *The Gauss curvature of a minimal surface is non-positive and unless the surface is a plane it can have only iso- lated zeros.*

Proof: By (8.8), S is a plane $\iff$ N constant $\iff$ g con- stant $\iff$ $g' \equiv 0$. But g' is analytic and either has isolated ze- ros or is identically zero. ◆

Let us note the important formula

(9.5) $$K = - \frac{\Delta \log \lambda}{\lambda^2}$$

which may be verified by a direct computation from (8.7) and (9.4). Actually it is proved in differential geometry that formula (9.5) holds for an arbitrary surface in terms of isothermal parameters, $g_{ij} = \lambda^2 \delta_{ij}$.

Consider now, for an arbitrary minimal surface in E^3, the follow- ing sequence of mappings, where Σ denotes the unit sphere:

(9.6) $D \xrightarrow{\ x(\zeta)\ } S \xrightarrow{\ \text{Gauss map}\ } \Sigma \xrightarrow[\text{projection}]{\text{stereographic}} w$-plane.

The composed map, as we have seen in the proof of Lemma 8.3, is

$$g(\zeta): D \to w\text{-plane.}$$

Consider an arbitrary differentiable curve $\zeta(t)$ in D, and its image under each of the maps in (9.6). If $s(t)$ is the arclength of the image on S, then as we have seen:

$$(9.7) \qquad \frac{ds}{dt} = \frac{1}{2} |f|(1 + |g|^2) \left| \frac{d\zeta}{dt} \right| .$$

The arclength of the image in the w-plane is simply

$$(9.8) \qquad \left| \frac{dw}{dt} \right| = |g'(\zeta)| \left| \frac{d\zeta}{dt} \right| .$$

If we let $\sigma(t)$ be arclength on the sphere, then it follows by the formula (6.1) for stereographic projection that

$$(9.9) \qquad \frac{d\sigma}{dt} = \frac{2}{1 + |w|^2} \left| \frac{dw}{dt} \right| .$$

Combining these formulas with (9.4), we find

$$(9.10) \qquad \frac{d\sigma}{dt} \Big/ \frac{ds}{dt} = \frac{4|g'|}{|f|(1+|g|^2)^2} = \sqrt{|K|} .$$

Thus we have an interpretation of the Gauss curvature in terms of the Gauss map. If we now let Δ be a domain whose closure is in D, the surface defined by the restriction of $x(\zeta)$ to Δ has *total curvature* given by

$$(9.11) \quad \iint_\Delta K \, dA = \iint_\Delta K \lambda^2 d\xi_1 d\xi_2 = -\iint_\Delta \left[\frac{2|g'|}{1 + |g|^2} \right]^2 d\xi_1 d\xi_2$$

which is the *negative of the area of the image of* Δ *under the Gauss map.* (This is in fact the original definition used by Gauss to define the Gauss curvature of an arbitrary surface.) Of course if the image is multiply-covered the total area is that of all the sheets. We may note that (9.11) may be regarded equivalently as the *spherical area of the image of* Δ *under* $g(\zeta)$.

Throughout this discussion we have been assuming that the surface S was defined in a plane domain D. However, the normal N and the value of the function g are independent of the choice of parameter. In fact, if we introduce new isothermal parameters by a conformal map $\zeta(\tilde{\zeta})$, then we have $\tilde{\phi}_k(\tilde{\zeta}) = \phi_k(\zeta) \dfrac{d\zeta}{d\tilde{\zeta}}$, $k = 1, 2, 3,$

and by its definition (8.3) the value of g is unchanged. Thus, if we are given an arbitrary minimal surface we may replace (9.6) by

$$(9.12) \qquad M \xrightarrow{\ x(p)\ } S \xrightarrow{\text{Gauss map}} \Sigma \xrightarrow[\text{projection}]{\text{stereographic}} w\text{-plane}$$

where the composed map

$$(9.13) \qquad\qquad g(p): M \to w\text{-plane}$$

is a meromorphic function on M. The total curvature of S is still the negative of the area of the image on Σ. In order to study the nature of this map for complete minimal surfaces, we prove a series of lemmas, of which the first is a generalization, and in a certain respect, a clarification of Lemma 8.5.

LEMMA 9.2. *Let* D *be a plane domain and* $g_{ij} = \lambda^2 \delta_{ij}$ *a Riemannian metric in* D, *where* $\lambda = \lambda(z) \in C^2$, *such that* D *is complete with respect to this metric. If there exists a harmonic function* $h(z)$ *in* D *such that*

$$(9.14) \qquad\qquad \log \lambda(z) \leq h(z)$$

throughout D, then D is either the entire plane, or else the plane with one point removed.

Proof: Introduce a second metric $\tilde{g}_{ij} = \tilde{\lambda}^2 \delta_{ij}$, where $\tilde{\lambda} = e^h$. Then $\lambda \leq \tilde{\lambda}$ throughout D, and if C is any divergent path in D, we have

$$\int_C \tilde{\lambda} |dz| \geq \int_C \lambda |dz| = \infty,$$

so that D is complete with respect to this metric also, and the same is true for the universal covering surface $\hat{D}$ of D with respect to the induced metric. But in a neighborhood of any point of $\hat{D}$ we may introduce an analytic function $f(z)$ whose real part is $h(z)$, and the mapping

$$w = \int e^{f(z)} dz$$

which satisfies

(9.15) $$\left| \frac{dw}{dz} \right| = \left| e^{f(z)} \right| = e^{h(z)} = \tilde{\lambda} \quad .$$

Thus the length of any curve on $\hat{D}$ with respect to the metric $\tilde{g}_{ij}$ is equal to the euclidean length of its image in the w-plane. By the simple connectivity of $\hat{D}$ there exists a global map of $\hat{D}$ into the w-plane satisfying (9.15), and by the completeness of $\hat{D}$, this map must be a one-to-one map of $\hat{D}$ onto the entire w-plane. Thus the universal covering surface of D is conformally equivalent to the plane, and it follows that D itself is of the type asserted in the lemma.

LEMMA 9.3. *Let D be the domain $0 < r_1 < |z| < r_2 \leq \infty$, and* *let* $g_{ij} = \lambda^2 \delta_{ij}$ *be a Riemannian metric in* D *satisfying* (9.14). *Suppose that* $\int_C \lambda |dz| = \infty$ *for every path* C *of the form* $z(t)$, $0 \leq t < 1$, *such that* $\lim_{t \to 1} |z(t)| = r_2$. *Then* $r_2 = \infty$.

Proof: Suppose $r_2 < \infty$. Then we may assume that

$$r_1 < \frac{1}{r_2} < 1 < r_2 \; ; \quad \Delta = \{z: \frac{1}{r_2} < |z| < r_2\} \; .$$

We introduce in Δ the metric $\tilde{g}_{ij} = \tilde{\lambda} \delta_{ij}$, where $\tilde{\lambda}(z) = \lambda(z) \lambda(1/z)$. Then one verifies easily that this metric satisfies the hypotheses of Lemma 9.2 in Δ, but the conclusion is not valid. Thus the assumption $r_2 < \infty$ leads to a contradiction. ◆

LEMMA 9.4. *Let* D *be a hyperbolic domain and* $g_{ij} = \lambda^2 \delta_{ij}$ *a metric with respect to which* D *is complete. Suppose that*

$$(9.16) \qquad\qquad \Delta \, \log \lambda \geq 0$$

$$(9.17) \qquad\qquad \iint_D |\Delta \, \log \lambda| \, dx \, dy < \infty, \qquad z = x + iy \; .$$

Then there exists a harmonic function $h(z)$ *in* D *satisfying* (9.14).

Proof: Since D is hyperbolic, for each point ζ in D there exists a Green's function $g(z, \zeta)$ which is harmonic and positive for $z \neq \zeta$, and such that $g(z, \zeta) + \log |z - \zeta| = H(z, \zeta)$, a harmonic function of z throughout D. (See, for example, Ahlfors and Sario [1], IV 6.) Set

$$u(\zeta) = \frac{1}{2\pi} \iint_D G(z, \zeta) \Delta \, \log \lambda(z) \, dx \, dy \; .$$

This integral exists, by (9.17) and the basic properties of the

Green's function, while $u(\zeta) \geq 0$ by (9.16). But by Poisson's formula, we have $\Delta u = -\Delta \log \lambda$, so that $h(z) = u + \log \lambda$ is harmonic in D. But since $u \geq 0$, $h(z) \geq \log \lambda$. ♦

REMARK. The formula (9.5), which holds for the Gauss curvature of an arbitrary surface in E^3, can also be used to define the Gauss curvature of an arbitrary Riemannian metric of the form $g_{ij} = \lambda^2 \delta_{ij}$. One can verify directly that it is invariant under conformal changes of parameters. Then (9.16) is equivalent to

$$(9.18) \qquad\qquad K \leq 0$$

and (9.17) is equivalent to

$$(9.19) \qquad\qquad \iint_D |K| \, dA < \infty \, .$$

THEOREM 9.1. *Let M be a complete Riemannian 2-manifold whose Gauss curvature satisfies (9.18) and (9.19). Then there exists a compact 2-manifold $\tilde{M}$, a finite number of points $p_1, \ldots, p_k$ on $\tilde{M}$, and an isometry between M and $\tilde{M} - \{p_1, \ldots, p_k\}$.*

Proof: By a theorem of A. Huber, the condition (9.19) on a complete 2-manifold M implies that M is finitely connected (A. Huber [1], p. 61). This means that there exists a relatively compact region M_0 on M bounded by a finite number of analytic Jordan curves $\gamma_1, \ldots, \gamma_k$, such that each component M_j of $M - M_0$ is doubly connected. (See Ahlfors and Sario [1], I 44 D and II 3 B.) Then each M_j can be mapped conformally onto an annulus $D_j : 1 < |z| < r_j \leq \infty$, where the curve γ_j corresponds to $|z| = 1$. The metric on M_j takes the form $g_{ij} = \lambda^2 \delta_{ij}$ in D_j, and the conditions (9.18) and (9.19) reduce to (9.16) and (9.17) respectively. The region D_j is obviously hyperbolic, since the function $\text{Re}\{\frac{1}{z} - 1\}$ is a negative harmonic function. By Lemma 9.4 there exists a harmonic function $h(z)$ satisfying (9.14) and by Lemma 9.3 (and the completeness of M) if follows that $r_j = \infty$. Let $\tilde{D}_j$ be the extension of D_j to a disk on the Riemann sphere obtained by adding the point at ∞. Let $\tilde{M}$ be the compact surface obtained by "welding" the disks $\tilde{D}_j$ to M_0 along γ_j (see Ahlfors and Sario [1], II 3C). Then M is conformally equivalent to $\tilde{M} - \{p_1, ..., p_k\}$, where p_j is the point $\tilde{D}_j - D_j$, and by carrying over the metric on M this correspondence becomes an isometry. ♦

LEMMA 9.5. *Let* $x(p): M \to E^3$ *define a complete regular minimal surface* S. *If the total curvature of* S *is finite, then the conclusion of Theorem 9.1 is valid, and the function* $g(p)$ *in (9.13) extends to a meromorphic function on* $\tilde{M}$.

Proof: We know that on a minimal surface $K \leq 0$, and therefore $|\iint K \, dA| = \iint |K| \, dA$, so that finite total curvature is equivalent to (9.19). Thus Theorem 9.1 may be applied, and we may consider $g(p)$ to be a meromorphic function on $\tilde{M} - \{p_1, ..., p_k\}$. If any of the points p_j were an essential singularity of g, then by Picard's theorem, g would assume every value infinitely often, with at most two exceptions. But this would imply that the spherical area of the image is infinite, and hence also the total curvature, contrary to assumption. Thus g has at most a pole at each p_j, and is therefore meromorphic on all of $\tilde{M}$. ♦

THEOREM 9.2. *Let* S *be a complete minimal surface in* E^3. *Then the total curvature of* S *can only take the values* $-4\pi m$, m *a non-negative integer, or* $-\infty$.

Proof: Since $K \leq 0$, either the integral $\iint K\,dA$ over the whole surface diverges to $-\infty$, or else the total curvature is finite. In the latter case we apply Lemma 9.5 and find that the total curvature is the negative of the spherical area of the image under g of $\tilde{M} - \{p_1, ..., p_k\}$. But since g is meromorphic on $\tilde{M}$, it is either constant, in which case $K \equiv 0$, or else it takes on each value a fixed number of times, say m. The spherical area of the image is then $4\pi m$. ◆

LEMMA 9.6. *Let $f(z)$ be analytic and different from zero for $0 < r_1 < |z| < \infty$. Suppose that for every path C which diverges to infinity, we have*

$$(9.20) \qquad \int_C |f(z)|\,|dz| = \infty \ .$$

Then $f(z)$ has at most a pole at infinity.

Proof: Since $f(z) \neq 0$, $\log|f(z)|$ is harmonic, and we have the Laurent expansion at infinity

$$\log|f(z)| = a \, \log|z| + h(z) + H(z) \ ,$$

where $h(z)$ is harmonic and bounded for $r_1 < r_2 < |z| < \infty$, and $H(z)$ is harmonic in the finite plane $|z| < \infty$. If N is any positive integer greater than a, it follows that for $|z| > r_2$,

$$(9.21) \qquad |f(z)| = |z|^a e^{h(z)} e^{H(z)} \leq M|z|^N e^{H(z)} = M|z^N e^{G(z)}| \ ,$$

where M is a suitable constant, and $G(z)$ is an entire function whose real part equals $H(z)$. We introduce the entire function

$$w = F(z) = \int z^N e^{G(z)} dz \ , \qquad F(0) = 0 \ ,$$

and note that there exists a single-valued branch

$$\zeta(z) = [F(z)]^{1/N+1}$$

in a neighborhood of $z = 0$, satisfying $\zeta'(0) \neq 0$. We therefore have
an inverse $z(\zeta)$ in a neighborhood of $\zeta = 0$, and either this inverse
extends to the whole ζ-plane, or else there is a largest circle
$|\zeta| < R$ to which it can be extended. But the latter cannot occur,
for then there would be a point ζ_0 satisfying $|\zeta_0| = R$ over which
the inverse could not be extended. If we consider the curve $\zeta(t) =$
$t\zeta_0$, $0 \leq t < 1$, its inverse image in the z-plane will be a path C
such that

$$\int_C |z^N e^{G(z)}| \, |dz| = \int_C |F'(z)| \, |dz| = R^{N+1} \quad .$$

If the path C were to diverge to infinity, then from some point on
it would be in the region $|z| > r_2$, and because of (9.21) this would
contradict (9.20). Thus C cannot diverge to infinity, and there
must exist a sequence of points z_n on C which have a finite point
of accumulation z_0, such that the images of these points in the
ζ-plane tend to ζ_0. But since $F'(z_0) \neq 0$, we could extend the inverse mapping over ζ_0. We thus conclude that the inverse $z(\zeta)$
is defined in the whole ζ-plane. Furthermore,

$$z(\zeta_1) = z(\zeta_2) \implies F(z(\zeta_1)) = F(z(\zeta_2)) \implies \zeta_1^{N+1} = \zeta_2^{N+1} \quad .$$

Thus each value is taken on at most $N+1$ times, and $z(\zeta)$ must
be a polynomial. But $z(\zeta) = 0 \implies \zeta = 0$, and hence $z(\zeta) = A\zeta^k$
for some integer k. Finally, $z'(0) = Ak\zeta^{k-1} \neq 0 \implies k = 1$, $A \neq 0$.
Thus $F(z) = (z/A)^{N+1}$, and $G(z)$ must be constant. The same is

then true of $H(z)$, so that $|f(z)| \le M_1 |z|^N$ near infinity, and $f(z)$ has at most a pole at infinity. ♦

THEOREM 9.3.* *Let S be a complete regular minimal surface in E^3. Then*

(9.22)
$$\iint_S K \, dA \le 2\pi(\chi - k) .$$

where χ is the Euler characteristic of S, and k is the number of boundary components.

Proof: If the integral on the left of (9.22) diverges to $-\infty$, the result holds trivially. Otherwise S has finite total curvature, and we may apply Theorem 9.1 and Lemma 9.5. By virtue of the relation (8.8) between the function g and the normals to the surface we may assume (after a preliminary rotation in space) that $g(p) \ne 0$, ∞ at the points p_j, and that the poles of $g(p)$ are all simple poles. In a neighborhood of any point of $M = \tilde{M} - \{p_1, \ldots, p_k\}$ we have the representation (8.2) in terms of isothermal parameters ζ. As we have noted earlier, under a conformal change of parameters $\zeta(\tilde{\zeta})$, the corresponding functions $\tilde{\phi}_k(\tilde{\zeta})$ satisfy $\tilde{\phi}_k(\tilde{\zeta}) = \phi_k(\zeta) d\zeta/d\tilde{\zeta}$, and similarly for $f(\zeta)$. This implies that the existence of a zero of ϕ_k or f at a point, as well as its order, is independent of the choice of local parameters. By Lemma 8.2, f must have a double zero at each of the simple poles of g, and f has no other zeros. Thus, if g is of order m on $\tilde{M}$, f has exactly $2m$ zeros on M. At each point p_j we may introduce local coordinates ζ so that $\zeta = 0$ corresponds to p_j. For $0 < |\zeta| < \epsilon$ we have $|g| < M$, hence $f \ne 0$. Furthermore, if C is a curve on M tending to p_j, we have by the completeness of S that

---

*See Appendix 3, Section 4.

$$\infty = \int_C \lambda |d\zeta| = \frac{1}{2} \int_C |f|(1 + |g|^2)|\zeta| \le \frac{1 + M^2}{2} \int_C |f| \, |d\zeta| \ .$$

It follows from Lemma 9.6 that f must have a pole at the origin. From the fact that the functions $x_k(\zeta)$ in (8.6) are single-valued in $0 < |\zeta| < \varepsilon$, and because of the assumption that $g(p_j) \ne 0$, it follows easily that the order ν_j of the pole of f at p_j is at least equal to 2. Thus $f(\zeta)d\zeta$ is a meromorphic differential on $\tilde{M}$, and one has the Riemann relation that the number of poles minus the number of zeros equals $2 - 2G$, where G is the genus of $\tilde{M}$ (Ahlfors and Sario [1], V 27 A). Furthermore, the Euler characteristic χ of M is equal to $2 - 2G - k$. We have, therefore

$$2 - 2G = \sum_{j=1}^{k} \nu_j - 2m \ge 2k - 2m$$

and

$$\iint_S K \, dA = -4\pi m \le 2\pi(2 - 2G - 2k) = 2\pi(\chi - k) \ . \qquad \blacklozenge$$

REMARK. The inequality $\int \int_S K \, dA \le 2\pi\chi$ was shown by Cohn-Vossen to hold for an arbitrary complete Riemannian 2-manifold with finite total curvature and finite χ. It follows in particular from (9.22) that in the case of minimal surfaces, equality can never hold in Cohn-Vossen's inequality. This question is discussed in a recent paper of Finn [6], who introduces at each boundary component a geometric quantity which acts as a compensating factor between the two sides of the Cohn-Vossen inequality. One obtains in this way a geometric interpretation of the order ν_j of the pole of f at p_j.

THEOREM 9.4. *There are only two complete regular minimal surfaces whose total curvature is* -4π. *These are the catenoid and Enneper's surface.*

Proof: This is the case $m = 1$ in Theorem 9.2. This means that the function g is meromorphic of order 1, hence maps $\tilde{M}$ conformally onto the Riemann sphere Σ. Thus $\tilde{M}$ has genus $G = 0$, and inequality (9.22) reduces to $k \leq 2$. We may therefore choose M to be Σ minus either one or two points, in which case we have $g(\zeta) = \zeta$, and by Lemma 9.6 $f(\zeta)$ is a rational function. Taking into account the completeness and the fact that the functions $x_k(\zeta)$ are single-valued, one easily finds that the only choices of $f(\zeta)$ compatible with these conditions yield precisely the two surfaces named.   ◆

COROLLARY. *Enneper's surface and the catenoid are the only two complete regular minimal surfaces whose Gauss map is one-to-one.*

Proof: If the Gauss map is one-to-one, then the total curvature, being the negative of the area of the image, satisfies

$$-4\pi \leq \iint_S K \, dA < 0 \ .$$

By Theorem 9.2, equality holds on the left, and the result follows from Theorem 9.4.   ◆

The methods used to prove the above theorems on the total curvature can also be used to study more precisely the behavior of the Gauss map, supplementing and sharpening the results obtained in the previous section. Let us give an example.

THEOREM 9.5. *Let* S *be a regular minimal surface, and suppose that all paths on the surface which tend to some isolated boundary component of* S *have infinite length. Then either the normals to* S *tend to a single limit at that boundary component, or else in each neighborhood of it the normals take on all directions except for at most a set of capacity zero.*

Proof: By a neighborhood of an isolated boundary component we mean a doubly-connected region whose relative boundary is a Jordan curve γ. This region on the surface may be represented by $x(\zeta): D \to E^3$, where D is an annular domain $1 < |\zeta| < r_2 \leq \infty$, the curve γ corresponding to $|\zeta| = 1$, and we may introduce the representation (8.2) in D. Suppose now that in some neighborhood of this boundary component the normals omit a set of positive capacity. This means that for some $r_1 \geq 1$, the function $w = g(\zeta)$ omits a set of positive capacity in the domain D': $r_1 < |\zeta| < r_2$. By Lemma 8.6 there exists a harmonic function $h(w)$ in the image of D' under $g(\zeta)$, such that $\log(1 + |w|^2) \leq h(w)$. Since the metric on S is given by $\lambda = \frac{1}{2}|f|(1 + |g|^2)$, we have

$$\log \lambda(\zeta) \leq \log \frac{|f(\zeta)|}{2} + \log h(g(\zeta)) \ .$$

Since the right-hand side is a harmonic function in D', we may apply Lemma 9.3 and deduce that $r_2 = \infty$. But then $g(\zeta)$ could not have an essential singularity at infinity by Picard's theorem. Thus $g(\zeta)$ tends to a limit, finite or infinite, as ζ tends to infinity, and the same is true of the surface normal. ◆

Without giving the details, let us note that further analysis along the above lines leads to the following result:

Let S be a complete minimal surface in E^3. Then

S has infinite total curvature $\Longleftrightarrow$ the normals to S take on all
   directions infinitely often with the exception of at most a set
   of logarithmic capacity zero;

S has finite non-zero total curvature $\Longleftrightarrow$ the normals to S take
   on all directions a finite number of times, omitting at most
   three directions;

S has zero total curvature $\Longleftrightarrow$ S is a plane.

This result, and the other theorems in this section are contained in the two papers Osserman [4, 5]. The presentation given here is somewhat different. Lemma 9.4 is based on an argument of A. Huber [1]. which he used to obtain the following result:

if S is a complete surface, and $\iint_S K^- dA$ converges, where
   $K^- = \max\{-K, 0\}$, then S is parabolic.

(This result was obtained earlier by Blanc and Fiala in the case that S is simply connected and the metric real analytic.) Lemma 9.6 was announced (without proof) by MacLane [1] and Voss [1]. The proof given here is taken from Finn [6] who obtains by the same reasoning a much more general result, not needed for our purposes. The corollary to Theorem 9.4 was observed independently by Osserman [6] and Voss [1].

In conclusion, let us note that it would be interesting in connection with the above results to have more examples of complete minimal surfaces whose genus is different from zero.*It was shown by Klotz and Sario [1] that there exist complete minimal surfaces of arbitrary genus and connectivity. Furthermore, it can be shown (Osserman [6]) that the classical surface of Scherk has infinite genus. There remains the following open question.

---

*See Appendix 3. Section 4.

Problem: *Does there exist a complete minimal surface of finite total curvature whose normals omit three directions?*

If so, the value "three" is the precise bound. If not, the maximum is "two," and is attained by the catenoid.

§10. *Non-parametric Minimal Surfaces in* E^3.

The study of minimal surfaces in non-parametric form consists of the study of solutions of the minimal surface equation

(10.1) $$(1 + q^2)r - 2pqs + (1 + p^2)t = 0,$$

and as such it may be considered as a chapter in the theory of non-linear elliptic partial differential equations. Just as we have seen in previous sections that the theory of functions of a complex variable and of Riemann surfaces led to many geometric results about minimal surfaces, we shall now show that many properties of minimal surfaces can be attributed to the form of equation (10.1). In the next section we shall reverse the process and use the parametric theory to derive properties of solutions of this equation.

The underlying idea in the present section is the comparison of an unknown solution of (10.1) with a fixed solution whose properties are well known. This method is possible by virtue of the following basic lemma.

LEMMA 10.1. *Let* F *and* G *be two solutions of equation (10.1) in a bounded domain* D. *Suppose that* $\lim(F - G) \leq M$ *for an arbitrary sequence of points approaching the boundary of* D. *Then* $F - G \leq M$ *throughout* D.

Proof: This lemma holds for a wide class of equations of which (10.1) is a special case. We refer to the discussion in Courant-Hilbert [1], p. 323.   ♦

An interesting feature of the study of equation (10.1) is that its solutions behave in some ways precisely as one would expect from the general theory of elliptic equations, as in the case of the above lemma, and in other ways they reveal completely unexpected properties. The non-existence of any solution in the whole plane other than the trivial linear ones (Bernstein's theorem) is an example of the latter. We shall now give some further examples. We first show that for certain domains D, the conclusion of Lemma 10.1 may be valid even though the hypotheses are known to hold on only part of the boundary. Let us introduce the following notation. Let

$$(10.2) \qquad G(r; r_1) = r_1 \cosh^{-1} \frac{r}{r_1}, \quad r \geq r_1; \quad G(r; r_1) \leq 0 .$$

The equation

$$(10.3) \qquad x_3 = G(r; r_1), \qquad r = \sqrt{x_1^2 + x_2^2} ,$$

defines the lower half of the catenoid, and is a solution of the minimal surface equation in the entire exterior of the circle $x_1^2 + x_2^2 = r_1^2$, taking the boundary values zero on this circle.

LEMMA 10.2. *Let D be a domain which lies in the annulus $r_1 < r < r_2$, and let $F(x_1, x_2)$ be an arbitrary solution of (10.1) in D. If the relation*

$$(10.4) \qquad \overline{\lim} \, (F(x_1, x_2) - G(r; r_1)) \leq M$$

is known to hold for an arbitrary sequence of points which tends to a boundary point of D not on the circle $r = r_1$, then

$$(10.5) \qquad F(x_1, x_2) \leq G(r; r_1) + M$$

throughout D.

REMARK. Perhaps the most striking illustration of this lemma is the case where D coincides with the annulus $r_1 < r < r_2$. Then the fact that (10.4) holds on the outside circle $r = r_2$ implies that (10.5) holds throughout D, and hence $\lim F(x_1, x_2) \leq M$ on the inner circle $r = r_1$. Geometrically, one can describe this situation as follows: given an arbitrary minimal surface lying over an annulus, if one places a catenoidal "cap" over the annulus in such a position that the surface lies below it along the outer circle, then the surface lies below it throughout the annulus. This behavior is in striking contrast with that of harmonic functions where on can prescribe arbitrarily large values on the inner circle, independently of the values on the outer circle, and find a harmonic function in the annulus taking on these boundary values.

Proof: For an arbitrary r_1' satisfying $r_1 < r_1' < r_2$, let

$$\varepsilon = \max_{r_1' \leq r \leq r_2} |G(r; r_1) - G(r; r_1')| \ .$$

Then by (10.4), we have

$$(10.6) \qquad \overline{\lim} \, (F(x_1, x_2) - G(r; r_1')) \leq M + \varepsilon$$

for all sequences approaching a boundary point of D lying in the annulus $r_1' \leq r \leq r_2$. We wish to conclude that

$$(10.7) \qquad F(x_1, x_2) \leq G(r; r_1') + M + \varepsilon$$

in the domain $D' = D \cap \{r_1' < r < r_2\}$. The result then follows by letting r_1' tend to r_1.

To prove (10.7), it suffices by Lemma 10.1 to show that (10.6) holds at every boundary point of D'. Since it already holds at boundary points of D, it suffices to show that (10.7) holds at interior points of D which lie on the circle $r = r_1'$. Suppose not.

Then the function $F(x_1, x_2) - G(r; r_1')$ would have a maximum $M_1 > M + \epsilon$ at some point (a_1, a_2) on this circle, interior to D. By Lemma 10.1, $F(x_1, x_2) - G(r; r_1') \leq M_1$ throughout D'. On the other hand, F has a finite gradient at the point (a_1, a_2), whereas by (10.2),

$$\frac{\partial G(r; r_1')}{\partial r}\Bigg|_{r=r_1'} = -\infty \ .$$

Since $F - G = M_1$ at (a_1, a_2), it follows that $F - G > M_1$ for all points in D' sufficiently near (a_1, a_2) and the origin. Thus the supposition that (10.7) did not hold leads to a contradiction, and the lemma is proved.   ♦

We shall give several applications of this lemma. The first is a generalized maximum principle for solutions of (10.1)

THEOREM 10.1. *Let* $F(x_1, x_2)$ *be a solution of the minimal surface equation* (10.1) *in a bounded domain D. Suppose that for every boundary point* (a_1, a_2) *of D, with the possible exception of a finite number of points, the relation*

(10.8)
$$\overline{\lim_{(x_1, x_2) \to (a_1, a_2)}} F(x_1, x_2) \leq M,$$

$$\underline{\lim_{(x_1, x_2) \to (a_1, a_2)}} F(x_1, x_2) \geq m$$

are known to hold. Then

(10.9) $$m \leq F(x_1, x_2) \leq M$$

throughout D.

Proof: It is sufficient to prove the right-hand inequality in (10.9), since the other then follows by considering the function $-F$.

We proceed by contradiction. Suppose that $\sup F = M_1 > M$. Then there exists a sequence of points in D along which F tends to M_1, and a subsequence of these points must converge to one of the exceptional boundary points. (If F assumes its maximum at an interior point of D, it must be constant, since x_3 is a harmonic function in local isothermal parameters.) We may suppose by a translation that this point is at the origin, and we may choose $r_2 > 0$ so that none of the other exceptional boundary points lies in the disk $r \le r_2$. Choose r_1 arbitrarily so that $0 < r_1 < r_2$ and let D' be the intersection of D with $\{r_1 < r < r_2\}$. Finally let $M_2 = \sup F$ for points of D on the circle $r = r_2$. We assert that $M_2 < M_1$. Namely, given a sequence of points on the circle $r = r_2$ along which F tends to M_2, they have a point of accumulation which either lies on the boundary of D, in which case $M_2 \le M$, or else interior to D, in which case $F < M_1$ at that point, since otherwise $F \equiv M_1 > M$ contradicting (10.8). Combining all these facts, we conclude that

$$\overline{\lim} \, (F(x_1, x_2) - G(r; r_1)) \le M_3 - G(r_2; r_1)$$

at all boundary points of D' for which $r > r_1$, where

$$M_3 = \max\{M, M_2\} < M_1 \ .$$

By Lemma 10.2, it follows that

$$F(x_1, x_2) \le G(r; r_1) - G(r_2; r_1) + M_3$$

throughout D'. Now for each fixed r, it follows from (10.2) that $G(r; r_1) \to 0$ as $r_1 \to 0$, and we deduce that $F(x_1, x_2) \le M_3 < M_1$ for $0 < r < r_2$. But this is in contradiction with $F \to M_1$ along a sequence of points tending to the origin, and the result follows. ♦

Let us note that in the statement of Theorem 10.1, each of the exceptional points may either lie on a boundary continuum, or else be an isolated boundary point. In the latter case the above proof simplifies considerably, and a much stronger result is in fact true.

THEOREM 10.2.[*] *A solution of the minimal surface equation cannot have an isolated singularity.*

Proof: Let $F(x_1, x_2)$ be an arbitrary solution of equation (10.1) in a punctured disk: $0 < r < \varepsilon$. We wish to show that $F(x_1, x_2)$ extends continuously to the origin, and that the resulting surface $x_3 = F(x_1, x_2)$ is a regular minimal surface over the whole disk $r < \varepsilon$.

Chose r_2, $0 < r_2 < \varepsilon$, and let $M = \max|F(x_1, x_2)|$ for $x_1^2 + x_2^2 = r_2^2$. By Theorem 10.1, $|F(x_1, x_2)| \leq M$ for $0 < x_1^2 + x_2^2 < r_2^2$. According to Theorem 7.2 there exists a function $\tilde{F}(x_1, x_2)$ defined and continuous in the disk $r \leq r_2$, satisfying (10.1) in $r < r_2$, and taking on the same values as F for $r = r_2$. Then $|\tilde{F}| \leq M$ also for $r \leq r_2$. We shall show that in fact $\tilde{F} = F$ for $0 < r < r_2$, and hence $\tilde{F}$ is the desired extension of the solution F. To this end we recall that in the notation

$$W = \sqrt{1 + p^2 + q^2}, \quad p = \frac{\partial F}{\partial x_1}, \quad q = \frac{\partial F}{\partial x_2},$$

the equation (10.1) can be written as

(10.10)
$$\frac{\partial \phi}{\partial x_1} + \frac{\partial \psi}{\partial x_2} = 0$$

where

(10.11)
$$\phi = \frac{p}{W}, \quad \psi = \frac{q}{W}.$$

*See Appendix 3, Section 5.

(This is merely the special case $n = 3$ of equation (3.14), but it can easily be verified directly.)

We now introduce the corresponding quantities for the function $\tilde{F}$, and consider the difference of these two functions in the annulus D_λ: $\lambda \leq r \leq r_2$. If we let C_λ be the circle $r = \lambda$, then by virtue of the fact that $F - \tilde{F} = 0$ on the circle $r = r_2$, Green's theorem applied to the domain D_λ yields

$$(10.12) \quad \int_{C_\lambda} (F - \tilde{F})[(\phi - \tilde{\phi})dx_2 - (\psi - \tilde{\psi})dx_1]$$

$$= \iint_{D_\lambda} [(p - \tilde{p})(\phi - \tilde{\phi}) + (q - \tilde{q})(\psi - \tilde{\psi})]dx_1 dx_2 ,$$

where we have applied equation (10.10) to both functions F and $\tilde{F}$. Now we note from (10.11) and the definition of W that $\phi^2 + \psi^2 < 1$ and similarly $\tilde{\phi}^2 + \tilde{\psi}^2 < 1$. Since $|F - \tilde{F}| \leq 2M$, it follows that the left-hand side of (10.12) tends to zero as $\lambda \to 0$. But the integrand on the right is non-negative, as one sees easily by applying Lemma 5.1 to the function $E(p, q) = \sqrt{1 + p^2 + q^2}$. We consider p and q as independent variables, and we find

$$\frac{\partial E}{\partial p} = \frac{p}{W} = \phi , \qquad \frac{\partial E}{\partial q} = \frac{q}{W} = \psi .$$

The integrand on the right of (10.12) then reduces to the expression on the left of (5.2). Thus if $(p - \tilde{p})^2 + (q - \tilde{q})^2 \neq 0$ at some point, it follows that the integrand on the right of (10.12) is positive at some point, hence in a neighborhood of that point, hence the integral could not tend to zero as λ tends to zero. Thus we conclude $p - \tilde{p} \equiv 0$, $q - \tilde{q} \equiv 0$ for $0 < r < r_1$, and hence also $F \quad \tilde{F} \equiv 0$, which proves the theorem. ♦

The proof just given, together with the proofs of Theorem 10.1 and Lemma 10.2 are due to Finn [5]. Lemma 10.2 is in fact a special case of Finn's lemma ([5], p. 139). Theorem 10.2 and the case of Theorem 10.1 in which the exceptional boundary points are isolated are originally due to Bers [1]. The form of Theorem 10.1 stated above is due to Nitsche and Nitsche [1]. It was later generalized to allow the exceptional set to be an arbitrary set of linear Hausdorff measure zero (Nitsche [5]). Similarly, Bers' theorem on removable singularities has the following extension:*If D is a bounded domain, and E a compact subset of D whose linear Hausdorff measure is zero, then every solution of the minimal surface equation in D − E extends to a solution in all of D. This result was obtained independently by Nitsche [5] and De Giorgi and Stampacchia [1]. The situation is quite different for the case $n > 3$. There, even bounded solutions of the minimal surface equation may have isolated non-removable singularities (Osserman [10]).†

We turn next to another application of Lemma 10.2. We wish to investigate the solvability of the Dirichlet problem for the minimal surface equation. For this purpose we introduce the following notation.

DEFINITION. Let D be a plane domain and P a boundary point of D. We say that P is a *point of concavity* of D if there exists a circle C through P and some neighborhood of P whose intersection with the exterior of C lies in D. The circle C is called a *circle of inner contact* at P.

LEMMA 10.3. *Let D be a bounded plane domain, P a point of concavity on the boundary, and C a circle of inner contact at P. If F is a solution of the minimal surface equation in D, then the*

---

*See Appendix 3, Section 5.
†See Appendix 3, Section 5.

boundary values of F at the point P are limited by the boundary values of F on the part of the boundary exterior to C.

Proof: We may assume that the circle C is centered at the origin and has radius r_1. The entire domain D is contained in some circle $r < r_2$. Suppose that $\overline{\lim}\, F \leq M$ for all boundary points of D exterior to C. Then by Lemma 10.2 applied to the intersection of D with the exterior of C, $F \leq G(r; r_1) - G(r_2; r_1) + M$ for $r_1 < r < r_2$, and therefore

$$(10.14) \qquad \lim_{Q \to P} F(Q) \leq M - G(r_2; r_1) \ . \qquad \blacklozenge$$

LEMMA 10.4. *Let D be an arbitrary plane domain. Then D is convex if and only if there does not exist a point of concavity on the boundary of D.*

Proof: Suppose first that P is a point of concavity of D, and let C be a circle of inner contact at P. Then a segment of the tangent line to C at P will have its endpoints in D, but the segment itself contains the point P not in D. Hence D is not convex. Conversely, if D is not convex there exist two points Q_1, Q_2 in D, such that the segment joining them is not in D. Let $Q(t)$, $0 \leq t \leq 1$ be a curve in D joining Q_1 to Q_2. Then there is a smallest value t_0 of t for which the segment L from Q_1 to $Q(t_0)$ is not entirely in D. We have $t_0 > 0$, and for all smaller values of t sufficiently near t_0 the line segments from Q_1 to $Q(t)$ lie in D and are on the same side of L. There is therefore an open subset Δ of D consisting of all points on one side of L and sufficiently near to L together with all points in a disk around each of the endpoints. By choosing a sufficiently flat circular arc lying in Δ

joining the endpoints of L and translating it until it first contracts the boundary of D, one finds a point of concavity.   ♦

THEOREM 10.3. *Let D be a bounded domain in the plane. Necessary and sufficient that there exist a solution of the minimal surface equation* (10.1) *in D taking on arbitrarily assigned continuous values on the boundary is that D be convex.*

Proof: Suppose first that D is convex. Then by Theorem 7.2 one can solve the boundary value problem for arbitrary continuous boundary values. If, on the other hand, D is not convex, then by Lemma 10.4 there exists a point of concavity P on the boundary of D. If we choose boundary values which are sufficiently large at P and small outside a neighborhood of P, then no solution can exist by virtue of the inequality (10.14).   ♦

This theorem and Lemma 10.3 are both due to Finn [5]. We may note that one can easily construct special cases of non-solvability directly from Lemma 10.2. For example, if D is the part of the annulus $r_1 < r < r_2$ lying in the first quadrant, and if we assign continuous boundary values which are equal to $G(r; r_1)$ outside the circle $r = r_1$, and positive somewhere on this circle, then no solution can exist by (10.5). The question of non-existence of solutions in non-convex domains had been considered earlier by a number of authors starting with Bernstein [2], but in each case the arguments had been valid only for special domains. For a detailed discussion of this question, see Nitsche [6].

We consider next a more general situation in which solutions may have infinite boundary values. We prove first the following result, also contained in the paper of Finn [5].

THEOREM 10.4. *Let D be an arbitrary domain having a point of concavity P on its boundary. Then a solution of the minimal surface equation in D cannot tend to infinity at P.*

Proof: Let C be a circle of inner contact at the point P. By choosing a slightly larger circle tangent to C at P, if necessary, we may assume that there is an arc γ of C which contains P but no other boundary points. If C' is a larger circle concentric with C and sufficiently close, then the region D' bounded by γ, two radial segments, and an arc γ' of C', will be entirely in D. Any solution F of (10.1) in D will have a finite upper bound M on the part of the boundary of D' exterior to C. We may then apply Lemma 10.3 to the domain D', and by (10.14) F cannot tend to infinity at the point P. ♦

The interest of the above theorem is that solutions of the minimal surface equation can indeed take on infinite boundary values, even along an entire arc of the boundary, as in the case of Scherk's surface

$$F(x_1, x_2) = \log \frac{\cos x_2}{\cos x_1}$$

which is defined in the square $|x_1| < \pi/2$, $|x_2| < \pi/2$, and tends to either $+\infty$ or $-\infty$ at every boundary point except for the vertices. Obviously if D is any domain interior to this square whose boundary is tangent to one of the sides, then F is a solution in D which tends to infinity at the point of tangency.

We next prove a result complementary to Theorem 10.4, which shows that infinite boundary values cannot occur along a whole convex arc.

THEOREM 10.5. *Let D be a plane domain, γ an arc of the boundary of D, L the line segment joining the endpoints of γ. If γ and L form the boundary of a subdomain D′ of D, then no solution of the minimal surface equation in D can tend to infinity at each point of γ.*

Proof: Let P be a point of $D′$ and let C be the circle through P and the endpoints of L. Then there exists a subdomain $D″$ of $D′$ which lies outside C and is bounded by parts of C and $γ$. If F were a solution of the minimal surface equation in D which approached infinity at each point of $γ$, then Lemma 10.2 applied to $-F$ would imply $F \equiv \infty$ in $D″$, which is impossible.   ♦

Combining Theorems 10.4 and 10.5, one sees the following:

if a solution of the minimal surface equation tends to infinity along a whole arc of the boundary, then that arc is a straight-line segment.

The question then arises whether one can solve the Dirichlet problem with infinite boundary values along a given line segment on the boundary. In the paper of Finn [4], which also contains Theorem 10.5, the following result is proved: *let D be a convex domain whose boundary contains a straight-line segment L. Let continuous boundary values be prescribed arbitrarily on the complementary part of the boundary. Then there exists a solution of the minimal surface equation in D which takes on these boundary values, and which tends to infinity at each inner point of L.*

A detailed study of the question of the Dirichlet problem with infinite boundary values has been made recently by H. Jenkins and J. Serrin [1, 2]. Necessary and sufficient conditions are given for the existence and uniqueness of solutions taking on prescribed

boundary values which are plus infinity, minus infinity, and continuous, respectively, on three given sets of boundary arcs. Let
us note one example of their results: *Let D be a convex quadri-
lateral. Let one pair of opposite sides have total length L, and
the other pair have total length M. Necessary and sufficient that
there exist a solution of the minimal surface equation in D taking
on the boundary values +∞ on one pair of sides and −∞ on the
other pair is that L = M. When the solution exists, it is unique up
to an additive constant.*

§11. *Application of parametric methods to non-parametric problems.*

We conclude our discussion of minimal surfaces in E^3 with several examples of how results on solutions of the minimal surface equation can be obtained by expressing the corresponding surfaces in parametric form.

THEOREM 11.1. *Let* $x_3 = F(x_1, x_2)$ *define a minimal surface in the disk* $D: x_1^2 + x_2^2 < R^2$. *Let* P *be the point on the surface lying over the origin, let* K *be the Gauss curvature of the surface at* P, *and let* d *be the distance along the surface from* P *to the boundary. Then the inequality*

(11.1)
$$|K| \leq \frac{c}{d^2 W_0^2}$$

holds, where c *is an absolute constant, and* W_0^2 *is the value at the origin of*

$$W^2 = 1 + \left(\frac{\partial F}{\partial x_1}\right)^2 + \left(\frac{\partial F}{\partial x_2}\right)^2 \quad .$$

REMARK. Inequality (11.1) may also be expressed in the form of a bound at the origin for the second derivatives of an arbitrary solution of (10.1) in the disk D.

Proof: We may represent the surface parametrically by $x(\zeta)$, $|\zeta| < 1$, the orientation being chosen so that the unit normal is

$$N = \left(\frac{p}{W}, \frac{q}{W}, -\frac{1}{W}\right) \; ; \qquad p = \frac{\partial F}{\partial x_1} \; , \qquad q = \frac{\partial F}{\partial x_2} \quad .$$

Using the representation (8.2), we have by (8.8) that

(11.2)
$$|g(\zeta)|^2 = \frac{W-1}{W+1} < 1 \quad .$$

If C is an arbitrary curve in $|\zeta| < 1$, then the length of its image on the surface is given, according to (8.7), by

$$\int_C \lambda |d\zeta| = \tfrac{1}{2} \int_C |f|(1 + |g|^2)|d\zeta| \leq \int_C |f| \, |d\zeta| \ .$$

In particular, if we consider all divergent paths C in $|\zeta| < 1$ which start at the origin, then we have

$$(11.3) \qquad d = \inf_C \int_C \lambda |d\zeta| \leq \int_{C_1} |f(\zeta)| \, |d\zeta| \ ,$$

where C_1 is any particular such path. We choose the path C_1 as follows. Set $w = G(\zeta) = \int f(\zeta) \, d\zeta$, where $G(0) = 0$. Then $G'(\zeta) = f(\zeta) \neq 0$ in $|\zeta| < 1$, by Lemma 8.2, and following the reasoning of Lemma 8.5, the inverse function $\zeta = H(w)$ defined in a neighborhood of the origin will be defined in a largest circle $|w| < \rho$ on whose boundary there will be a point w_0 over which the inverse cannot be extended. The radius $L: w(t) = tw_0$, $0 \leq t < 1$, will map onto a divergent path C_1 in $|\zeta| < 1$ which starts at the origin. But

$$(11.4) \qquad \int_{C_1} |f(\zeta)| \, |d\zeta| = \int_L |dw| = \rho \ .$$

By Schwarz' lemma, applied to the map $H(w)$ of $|w| < \rho$ into $|\zeta| < 1$, we have $|H'(0)| \leq 1/\rho$. Thus $|f(0)| = |G'(0)| = 1/|H'(0)| \geq \rho$. Comparing with (11.3) and (11.4) yields

$$(11.5) \qquad\qquad\qquad d \leq |f(0)| \ .$$

Next we apply Schwarz' lemma to the map $g(\zeta)$ of $|\zeta| < 1$ into $|g(\zeta)| < 1$, and we find

(11.6) $$|g'(0)| \leq 1 - |g(0)|^2 .$$

Finally, combining (11.2), (11.5) and (11.6) with the expression (9.4) for the Gauss curvature, we find

$$\sqrt{|K|}\, d \leq \frac{4|g'(0)|}{(1+|g(0)|^2)^2} \leq \frac{4(1-|g(0)|^2)}{(1+|g(0)|^2)^2} = \frac{2}{W_0}\left(1+\frac{1}{W_0}\right) \leq \frac{4}{W_0}$$

which proves the theorem. ♦

COROLLARY.* *Under the same hypotheses, there exists an absolute constant* c_0 *such that*

(11.7) $$|K| \leq \frac{c_0}{R^2}$$

Proof: Obviously $d \geq R$, and $W_0 \geq 1$. ♦

Let us note that the inequalities (11.1) and (11.7) may be regarded as quantitative forms of Bernstein's theorem. Namely, if $F(x_1, x_2)$ is defined over the whole (x_1, x_2) plane, then we may choose a disk of arbitrarily large radius R about any point, and it follows from (11.7) that the Gauss curvature of such a surface is identically zero, which for a minimal surface implies that it is a plane. The first results of this type were given by Heinz [1], where one finds inequality (11.7). This was later sharpened by E. Hopf [1], who gave a different proof and found that one could insert the factor W^2 in the denominator.

The inequality (11.1) is in Osserman [2]. It has the advantage that it remains true if the disk D is replaced by an arbitrary domain in the (x_1, x_2)-plane. Concerning the constant c, one can show that if the surface has a horizontal tangent plane at the origin, then $W_0 = 1$, and

---

*See Appendix 3, Sections 2, 6IB, and 6IIA.

$$|K| \leq \frac{64}{9} \frac{1}{d^2} \quad .$$

Furthermore, this inequality is sharp, equality being attained in the case of Enneper's surface (the domain D no longer being a disk in this case). By a refinement of the argument given above one can show that the constant c in (11.1) satisfies

$$\frac{64}{9} \leq c < 8 \quad .$$

These results are contained in the paper Osserman [2].

A completely different proof of inequality (11.7), using purely parametric methods is given in Finn and Osserman [1]. It is shown there that if the surface has a horizontal tangent plane at the origin, then one has

$$|K| < \frac{\pi^2}{2} \frac{1}{R^2} \quad ,$$

and that this inequality is best possible, with Sherk's surface playing the role of an extremal. It is also shown that the constant c_0 in (11.7) satisfies

$$\frac{\pi^2}{2} \leq c_0 < 6 \quad .$$

These considerations point obviously to the following question.

PROBLEM. What are the precise values of the constants c and c_0 in (11.1) and (11.7)? For a given value of W can one describe the extremal surface for these inequalities?

We turn next to solutions of the minimal surface equation defined in the *exterior* of a disk. We note first the following result, due to Bers [1].

THEOREM 11.2. *Let* $x_3 = F(x_1, x_2)$ *define a minimal surface in the region* $x_1^2 + x_2^2 > R^2$. *Then the gradient of* F *tends to a limit at infinity.*

Proof: The normals to such a surface are contained in a hemisphere, whereas the length of all curves on the surface which diverge to infinity is clearly infinite. It follows by Theorem 9.5 that the normals tend to a limit at infinity, which is the desired result. ◆

Actually, Bers obtained much more precise results on the behavior of the solution $F(x_1, x_2)$ in the neighborhood of infinity. For this purpose he used a parametric representation analogous to (8.2), together with the following lemma.

LEMMA 11.1. *Let* $x_3 = F(x_1, x_2)$ *define a minimal surface* S *in the region* $x_1^2 + x_2^2 > R^2$. *Then* S *is conformally equivalent to the punctured disk:* $0 < |\zeta| < 1$, *where* ζ *tends to zero as* $x_1^2 + x_2^2$ *tends to infinity.*

Proof: This result follows by the same reasoning that was used in the proof of Theorem 9.5.   ◆

The precise conditions in parametric form for a surface to be of the kind we wish to consider are the following.

LEMMA 11.2. *Let a minimal surface* S *be represented by the equations*

$$(11.8) \qquad x_k = \text{Re} \int \phi_k(\zeta) \, d\zeta$$

in the punctured disk $\Delta: 0 < |\zeta| < 1$, *where*

(11.9) $\phi_1 = \frac{1}{2} f(1-g^2)$, $\phi_2 = \frac{i}{2} f(1+g^2)$, $\phi_3 = fg$.

Then the following statements are equivalent:

   I. *There exists $\rho > 0$ such that the part of S corresponding to $0 < |\zeta| < \rho$ can be solved in the form $x_3 = F(x_1, x_2)$, where $F(x_1, x_2)$ is a bounded solution of the minimal surface equation in the exterior of some Jordan curve Γ.*

   II. *There exists $\rho' > 0$ such that in the punctured disk Δ': $0 < |\zeta| < \rho'$, the functions $f(\zeta)$, $g(\zeta)$ are analytic and have the following properties:*

      a) *$g(\zeta)$ extends to an analytic function in $|\zeta| < \rho'$, satisfying $|g(\zeta)| < 1$, $g(0) = 0$, $g'(0) = 0$.*
      b) *$f(\zeta) \neq 0$ in Δ', and $f(\zeta)$ has a pole of order two at the origin with residue zero.*

   Proof: I $\Longrightarrow$ II. Let $\rho' = \rho$. A surface in non-parametric form is necessarily regular, and satisfies $|g| < 1$, hence $f \neq 0$, by Lemma 8.2. Since g is a bounded analytic function in a punctured disk, it has a removable singularity at the origin. The value of $g(0)$ determines the limiting position of the gradient at infinity; if $g(0) \neq 0$, then by (11.2) the gradient would tend to a limit different from zero and $F(x_1, x_2)$ would not be bounded. Thus $g(0) = 0$. Furthermore, $x_3 = F(x_1, x_2)$ bounded at infinity means that x_3 is a bounded harmonic function of ζ in a neighborhood of the origin, and hence also has a removable singularity. Thus

$$f(\zeta) g(\zeta) = \phi_3(\zeta) = \frac{\partial x_3}{\partial \xi_1} - i \frac{\partial x_3}{\partial \xi_2} , \qquad \zeta = \xi_1 + i\xi_2 ,$$

is analytic in the full disk, and the same is true of the function

$f(\zeta) g(\zeta)^2$. From this it follows first of all that f cannot have an essential singularity at the origin. Further, from (11.8) and (11.9), we have

$$(11.10) \qquad x_1 + ix_2 = \frac{1}{2} \left[\overline{\int fd\zeta} - \int fg^2 d\zeta \right] ,$$

from which we deduce that f must have residue zero at the origin in order for x_1 and x_2 to be single-valued functions of ζ. On the other hand, f cannot be regular at the origin, since otherwise x_1 and x_2 would be bounded. Thus f has a pole of order at least two, and fg analytic at the origin implies g has a double zero; i.e., $g'(0) = 0$. Finally, we must show that $f(\zeta)$ cannot have a pole of order greater than two. But note that by (11.10) we may write

$$(11.11) \qquad x_1 + ix_2 = \frac{1}{2} [\overline{H(\zeta)} - G(\zeta)]$$

where

$$(11.12) \qquad H(\zeta) = \int f(\zeta) d\zeta , \qquad G(\zeta) = \int f(\zeta) g(\zeta)^2 d\zeta$$

are single-valued analytic functions in Δ'. If $f(\zeta)$ had a pole of order greater than two, then $H(\zeta)$ would have a pole of order greater than one, and the image of $|\zeta| = \varepsilon$ for all small ε would wind around the origin more than once. But $|H(\zeta)| > |G(\zeta)|$ on $|\zeta| = \varepsilon$ for small ε, and thus the image of $|\zeta| = \varepsilon$ in the x_1, x_2-plane would also wind around the origin more than once. But this contradicts the fact that for $0 < |\zeta| < \rho$, the map $\zeta \to (x_1, x_2)$ ia one-to-one, and the result is proved.

II $\Longrightarrow$ I. Conditions a) and b) guarantee that the functions $G(\zeta)$, $H(\zeta)$ defined by (11.12) are single-valued in Δ', and that $G(\zeta)$ is analytic at the origin, while $H(\zeta)$ has a simple pole. Thus, by (11.11)

$x_1^2 + x_2^2$ tends to infinity as ζ tends to zero. For the Jacobian of
the map (11.11) we have the expression

$$(11.13) \qquad J = \frac{\partial(x_1, x_2)}{\partial(\xi_1, \xi_2)} = \text{Im}\{\phi_1 \bar{\phi}_2\} = \tfrac{1}{4} |f|^2(|g|^4 - 1) .$$

Since $|g| < 1$, we have $J < 0$ everywhere, and the map (11.11) is
a local diffeomorphism. We wish to show that it is a global diffeo-
morphism in $0 < |\zeta| < \rho$, for suitable choice of ρ. To this end, we
first choose $r > 0$ so that $H(\zeta)$ is univalent in $0 < |\zeta| \le r$, which
is possible since $H(\zeta)$ has a simple pole. Let $M = \max |G(\zeta)|$ for
$|\zeta| \le r$, and choose $r_1 \le r$ so that $|H(\zeta)| > M$ for $0 < |\zeta| < r_1$.
Let $M_1 = \max|x_1 + ix_2|$ for $|\zeta| = r_1$. We assert that every point
in the exterior of the circle $x_1^2 + x_2^2 = M_1^2$ is taken on precisely
once in $0 < |\zeta| < r_1$. Namely, given any such point (a_1, a_2),
choose $r_2 < r_1$ such that $|H(\zeta)| > 2|a_1 + ia_2| + M = M_2$ for
$|\zeta| \le r_2$. Then $|x_1 + ix_2| > |a_1 + ia_2|$ on $|\zeta| = r_2$, and the map
(11.11) is a local diffeomorphism of the annulus $r_2 < |\zeta| < r_1$,
such that the image of the boundary has winding number equal to
one with respect to the point (a_1, a_2). The point (a_1, a_2) has
therefore exactly one pre-image in this annulus and none for
$|\zeta| \le r_2$. This completes the proof of the lemma. ◆

LEMMA 11.3. *Let* $F(x_1, x_2)$ *satisfy the minimal surface equa-*
tion in the exterior of some circle $x_1^2 + x_2^2 = R^2$. *If* $F(x_1, x_2)$ *is*
bounded, then it tends to a limit at infinity, and

$$(11.14) \qquad \left| \frac{\partial F}{\partial x_1} + i \frac{\partial F}{\partial x_2} \right| \le \frac{M}{x_1^2 + x_2^2}$$

in a neighborhood of infinity, for a suitable constant M.

Proof: By Lemmas 11.1 and 11.2 we have the representation (11.8), (11.9), where f and g satisfy conditions II a) and b) of Lemma 11.2. As we pointed out in the proof of Lemma 11.2, x_3 tends to a limit as ζ tends to zero, hence as $x_1^2 + x_2^2$ tends to infinity. From the representation (11.11) and the properties of G and H, it follows immediately that $|x_1 + ix_2| \, |\zeta|$ tends to a finite limit as ζ tends to zero. Finally, from (8.8) one may compute

$$\frac{\partial F}{\partial x_1} = \frac{2 \, \text{Re}\{g\}}{1 - |g|^2} \quad , \quad \frac{\partial F}{\partial x_2} = \frac{2 \, \text{Im}\{g\}}{1 - |g|^2} \quad ,$$

and since $g(\zeta)$ has a double zero at the origin,

$$\left| \frac{\partial F}{\partial x_1} + i \frac{\partial F}{\partial x_2} \right| = \frac{2|g|}{1 - |g|^2} \leq M|\zeta|^2 \leq \frac{M_1}{|x_1 + ix_2|^2}$$

in a neighborhood of infinity. ♦

THEOREM 11.3. *If the exterior Dirichlet problem has a bounded solution, then it is unique. Specifically, if $F(x_1, x_2)$, $\tilde{F}(x_1, x_2)$ are bounded solutions of the minimal surface equation in the domain $x_1^2 + x_2^2 > R^2$, if they are continuous for $x_1^2 + x_2^2 \geq R^2$ and agree for $x_1^2 + x_2^2 = R^2$, then $F(x_1, x_2) \equiv \tilde{F}(x_1, x_2)$.*

Proof: Let us use the same notation as in the proof of Theorem 10.2, but applied to the annulus $D: r_1 \leq r \leq r_2$, where $r_1 > R$. We obtain formula (10.12), but with two boundary integrals on the left. The one around the outside boundary tends to zero as r_2 tends to infinity by virtue of (11.4) and the fact that both F and F are bounded. The one on the inside circle tends to zero as $r_1 \to R$ because $F - \tilde{F}$ tends to zero, whereas the quantities $\phi, \tilde{\phi}, \psi, \tilde{\psi}$ are all less then 1 in absolute value. Thus the reasoning of Theorem 10.2 shows that F and $\tilde{F}$ have the same gradient, hence differ by at most a constant, and and since they coincide on a circle they are identical. ♦

REMARK. The above proof was suggested by R. Finn. Obviously the boundary need not be a circle, since the reasoning is quite general. The fact that this theorem was true was pointed out to the author by D. Gilbarg, who also raised the question of whether there always exists a bounded solution of the exterior Dirichlet problem. We show that there does not in the following theorem.

THEOREM 11.4. *There exist continuous functions on the circle* $x_1^2 + x_2^2 = 1$ *which may be chosen to be arbitrarily smooth (e.g., real analytic functions of arc length) such that no bounded solution of the minimal surface equation in the exterior of this circle takes on these values on the boundary.*

Proof: We first construct a particular surface $\tilde{S}$ which will play a role similar to that of the catenoid in the arguments of the previous section. This surface is defined by setting

$$f(\zeta) = \frac{2}{\zeta^2} \quad , \quad g(\zeta) = \zeta^2$$

in (11.9), for $0 < |\zeta| \leq 1$. The transformation (11.11) takes the form

(11.15) $x_1 + ix_2 = -1/\zeta - \zeta^3/3$.

One can easily verify that the image of $|\zeta| = 1$ under (11.15) is a Jordan curve Γ, and it follows by the same reasoning as in the proof of Lemma 11.2 that (11.15) is a diffeomorphism of $0 < |\zeta| < 1$ onto the exterior of Γ. Thus the surface $\tilde{S}$ can be expressed in non-parametric form $x_3 = \tilde{F}(x_1, x_2)$ in the exterior of Γ. Using the equation $x_3 = 2\xi_1$, one may easily verify the following facts:

   i) the map (11.15) takes the axes into the axes and symmetric points into symmetric points;

ii) $\tilde{F}(x_1, x_2)$ is symmetric with respect to the x_1-axis, and anti-symmetric with respect to the x_2-axis; it is positive for $x_1 < 0$, negative for $x_1 > 0$, and tends to zero at infinity;

iii) Γ may be expressed in polar form: $r = h(\theta)$, where r decreases monotonically from $4/3$ to $2/3$ as θ increases from 0 to $\pi/2$ (the rest of Γ being obtained by symmetry);

vi) the gradient of $\tilde{F}$ is infinite along Γ .

We now let C be the unit circle $x_1^2 + x_2^2 = 1$, and we assign boundary values $\phi(x_1, x_2)$ on C satisfying

a) ϕ is symmetric with the respect to the x_1-axis, antisymmetric with respect to the x_2-axis.

b) for points on C outside Γ in the left half-plane $x_1 < 0$, we have $0 < \phi(x_1, x_2) < \tilde{F}(x_1, x_2)$.

Suppose now that $F(x_1, x_2)$ were a bounded solution of the minimal surface equation in the exterior of C, taking on the boundary values $\phi(x_1, x_2)$. Then by property a) of ϕ, the function $\hat{F}(x_1, x_2) = -F(-x_1, x_2)$ would be a bounded solution taking on the same boundary values, and by Theorem 11.3, $F(x_1, x_2) \equiv \hat{F}(x_1, x_2)$. In particular $F(0, x_2) = \hat{F}(0, x_2) = -F(0, x_2)$, so that $F(0, x_2) \equiv 0$. Hence the limit of F at infinity must be zero. Thus, for any $\varepsilon > 0$, we can choose R sufficiently large so that $|F| < \varepsilon$, $|\tilde{F}| < \varepsilon$ on $x_1^2 + x_2^2 = R^2$. We now let D be the domain in the left half-plane, lying inside $x_1^2 + x_2^2 = R^2$, and outside the curves C and Γ. On the vertical axis we have $F = \tilde{F} = 0$. On $x_1^2 + x_2^2 = R^2$, we have $F < \tilde{F} + 2\varepsilon$, and for points on C outside Γ, $F = \phi < \tilde{F}$. Thus the equality $F < \tilde{F} + 2\varepsilon$ holds at all boundary points of D except possibly for those on Γ. However, along Γ the gradient of $\tilde{F}$ is infinite, and by the same reasoning as was used in

Lemma 10.2, the inequality $F < \bar{F} + 2\varepsilon$ must also hold for boundary points on Γ. Finally, if we let D' be the crescent-shaped domain in the left half-plane lying inside Γ and outside the circle C, then the inequality $F < \bar{F} + 2\varepsilon$ on Γ implies, again by Lemma 10.2, a bound of the form $F \leq M$ on the part of the boundary of D' on the circle C. This means that if the boundary values $\phi(x_1, x_2)$, which have not yet been prescribed on the part of C interior to Γ, are chosen so that $\phi > M$ at some point, then the exterior boundary problem cannot be solved. This proves the theorem. ◆

We conclude this discussion with the following remarks.

First of all, one could remove the requirement of boundedness and ask if there exists *any* solution of the exterior Dirichlet problem. However, one would then have to sacrifice uniqueness, since even for constant boundary values on the circle one has distinct solutions in the catenoid and a horizontal plane.

Next we note that the choice of a circle as boundary curve is of no special importance. Obviously the argument works in great generality. It would be interesting to give an argument for an arbitrary exterior domain.

Finally, we note that according to an observation of D. Gilbarg, one can prove by standard methods in the theory of uniformly elliptic equations that a solution of the exterior Dirichlet problem will always exist if the boundary data are sufficiently small. We thus have, by virtue of the non-existence for certain sufficiently large boundary values, one more striking example of how the non-linearity of the minimal surface equation affects the behavior of its solutions.

§12. *Parametric surfaces in E^n; generalized Gauss map.*

In this section we shall show how the results of sections 8 and 9 may be extended to minimal surfaces in E^n. For this purpose we must define the generalization to n dimensions of the classical Gauss map, and study its properties in the case of minimal surfaces. This was first done for $n = 4$ by Pinl [1]. We may also note here that Pinl has obtained a number of results on parametric minimal surfaces in E^n which we are not able to discuss here. (See Pinl [2, 3] and further references given there. Note also the papers of Beckenbach [1] and Jonker [1].) For arbitrary n, the Gauss map of minimal surfaces was first studied by Chern [1]. The present section is devoted to the results in that paper, together with those given in Osserman [5] and in Chern and Osserman [1].

We begin by discussing the Grassmannian $G_{2,n}$. This is a differentiable manifold whose points are in one-to-one correspondence with the set of all oriented two-dimensional linear subspaces of E^n; i.e., all oriented planes through the origin. Let Π be any such plane, and let v, w be a pair of orthogonal vectors of equal length which span Π, and such that the ordering v, w agrees with the orientation of Π. We then set $z_k = v_k + iw_k$, $k = 1, \ldots, n$, and we thus obtain a map

(12.1) $$v, w \to z \in \mathbf{C}^n .$$

Suppose next that v', w' are another such pair of vectors spanning Π, and z' is their image under the map (12.1). Then one verifies easily that there exists a complex constant c such that $z_k' = c\, z_k$ for all k. Thus the map (12.1) induces a map

(12.2) $$\Pi \to z \in P^{n-1}(\mathbf{C})$$

which assigns to each plane Π a point in the $(n-1)$-dimensional complex projective space. Furthermore, we have

$$\sum_{k=1}^{n} z_k^2 = \sum v_k^2 - \sum w_k^2 - 2i \sum v_k w_k = 0.$$

Thus, if we introduce the complex hyperquadric

$$(12.3) \qquad Q_{n-2} = \left\{ z \, \epsilon \, P^{n-1}(\mathbf{C}) : \sum_{k=1}^{n} z_k^2 = 0 \right\},$$

the map (12.2) actually takes the form

$$(12.4) \qquad \Pi \to z \, \epsilon \, Q_{n-2} \subset P^{n-1}(\mathbf{C}).$$

Conversely, for any point $z \, \epsilon \, Q_{n-2}$, if we write $z_k = v_k + i w_k$, then the vectors v, w will span a plane Π which corresponds to the point z. We thus see that the map (12.4) defines a one-to-one correspondence between all planes Π and all points of Q_{n-2}. We thus identify Q_{n-2} with the Grassmannian $G_{2,n}$.

Suppose now that we have a regular C^r-surface S defined by a map

$$(12.5) \qquad x(p) : M \to E^n,$$

where M is an oriented 2-manifold. Then at each point we have a tangent plane $\Pi(p)$. This may be defined in terms of local parameters u_1, u_2 as the plane spanned by the vectors $\partial x / \partial u_1$, $\partial x / \partial u_2$. and is independent of choice of parameters. We may then assign the point $z(p)$ obtained by the map (12.4), and the resulting map

$$(12.6) \qquad g : S \to Q_{n-2}$$

where

$$(12.7) \qquad g : x(p) \longmapsto \Pi(p) \longleftrightarrow z(p)$$

is called the *generalized Gauss map* of the surface S. In the case $n = 3$, we have a one-to-one correspondence between oriented planes Π and their unit normals N, and one-to-one correspondence between the points of Q_1 and those of the unit sphere obtained by setting $w = z_3/(z_1 - iz_2)$ and following by stereographic projection. The map (12.7) then reduces to the classical Gauss map: $x(p) \to N(p)$. We will simply refer to the map (12.7) as the "Gauss map" for arbitrary n.

We recall next that the immersion (12.5) defines a natural conformal structure on M, where local parameters on M are those which yield isothermal parameters on S.

LEMMA 12.1. *A surface S defined by (12.5) is a minimal surface if and only if the Gauss map (12.7) is anti-analytic.*

Proof:[*] In a neighborhood of any point the surface S may be represented by $x(\zeta)$ in terms of isothermal parameters $\zeta = \xi_1 + i\xi_2$. Then the vectors $\partial x/\partial \xi_1$, $\partial x/\partial \xi_2$ are orthogonal, equal in length and span the tangent plane. Thus we may define the Gauss map by setting

$$(12.8) \qquad z_k = \frac{\partial x_k}{\partial \xi_1} + i\frac{\partial x_k}{\partial \xi_2} = \overline{\phi_k(\zeta)} \quad .$$

By Lemma 4.2 we have:

S a minimal surface $\Longleftrightarrow$ x_k harmonic $\Longleftrightarrow$ $\overline{z}_k$ analytic in ζ. ♦

LEMMA 12.2. *A surface S in E^n lies in a plane if and only if its image under the Gauss map reduces to a point.*

Proof: The image under the Gauss map is a point if and only if the tangent vectors $\partial x/\partial u_1$, $\partial x/\partial u_2$ may be expressed as linear

---

[*]This proof is not complete as it stands. For a discussion and complete proof, see pp. 7–10 of Hoffman and Osserman [1].

combinations of two fixed vectors v, w, which is equivalent to $x(p) - x(p_0)$ expressible as a linear combination of v and w for all points p and a fixed point p_0. ◆

THEOREM 12.1.* *Let S be a complete regular minimal surface in E^n. Then either S is a plane, or else the image of S under the Gauss map approaches arbitrarily closely every hyperplane in $P^{n-1}(\mathbf{C})$.*

Proof: We may first of all pass to the universal covering surface $\hat{S}$ of S, which affects neither the hypothesis nor the conclusion. Suppose that the image under the Gauss map does not come arbitrarily close to a certain hyperplane $\Sigma \, a_k z_k = 0$. This means that

$$(12.9) \qquad \frac{|\sum_{k=1}^{n} a_k z_k|^2}{\sum_{k=1}^{n} |z_k|^2} \geq \epsilon > 0$$

everywhere on the image.

Let $\hat{S}$ be defined by $x(\zeta): D \to E^n$, where D may be either the unit disk or the whole ζ-plane. Then in the notation (12.8), the inequality (12.9) takes the form

$$(12.10) \qquad \sum_{k=1}^{n} |\phi_k(\zeta)|^2 \leq \frac{1}{\epsilon} |\psi(\zeta)|^2$$

where $\psi(\zeta) = \Sigma \, \bar{a}_k \phi_k(\zeta)$ is analytic and different from zero in D. But the metric on $\hat{S}$ is given by $g_{ij} = \lambda^2 \delta_{ij}$, where $\lambda^2 = \frac{1}{2} \Sigma |\phi_k(\zeta)|^2$, and by (12.10) $\log \lambda$ has a harmonic majorant in D. Since $\hat{S}$ is complete, it follows from Lemma 9.2 that D cannot be the unit circle, but must be the entire plane. But in that case, it follows from

---

*See Appendix 3, Section 4.

(12.10) that each of the functions $\phi_k(\zeta)/\psi(\zeta)$ is a bounded entire function, hence constant. Thus the image of $\hat{S}$ under the Gauss map reduces to a point, and by Lemma 12.2 $\hat{S}$ lies in a plane, and being complete, $\hat{S}$ coincides with this plane. ♦

COROLLARY. *The normals to a complete regular minimal surface S in E^n are everywhere dense unless S is a plane.*

Proof: Suppose that the normals to S all make an angle of at least α with a certain unit vector $(\lambda_1, ..., \lambda_n)$. One can show that this is equivalent to the inequality

$$(12.11) \qquad \frac{|\Sigma \lambda_k z_k|^2}{\Sigma |z_k|^2} \geq \frac{1}{2} \sin^2 \alpha ,$$

which means that the image of S under the Gauss map is bounded away from the hyperplane $\Sigma \lambda_k z_k = 0$. The result therefore follows from Theorem 12.1. To obtain (12.11) one need only choose a pair of orthogonal unit vectors v, w which span the tangent plane at a point, and decompose λ in the form $\lambda = av + bw + N$, where N is a normal at the point. If ω is the angle between N and λ, then since $\lambda \cdot N = |N|^2$, we have

$$(\lambda \cdot v)^2 + (\lambda \cdot w)^2 = a^2 + b^2 = 1 - |N|^2 = 1 - \cos^2 \omega \geq 1 - \cos^2 \alpha$$

which is precisely (12.11), where $z_k = v_k + i w_k$. ♦

The above theorem and its corollary may be regarded as the first steps in the direction of settling the following basic problems:

1. *Describe the behavior of the Gauss map for an arbitrary complete minimal surface in E^n.*

2. *Relate this behavior to geometric properties of the surface itself.*

In sections 8 and 9 we have given a number of results of this type for the case $n = 3$. We now review briefly the situation for arbitrary n. First let us make the following definition.

DEFINITION. The Gauss map is *degenerate* if the image lies in a hyperplane.

LEMMA 12.3. *Let S be a minimal surface in E^n whose image under the Gauss map lies in a hyperplane*

$$(12.12) \qquad H: \Sigma\, a_k z_k = 0\,.$$

If H is a "real" hyperplane; i.e., the a_k are real, then S itself lies in a hyperplane of E^n.

If H is a tangent hyperplane to Q_{n-2}, i.e.,

$$(12.13) \qquad \Sigma\, a_k^2 = 0,$$

then there exists a change of coordinates in E^n, say $x_j = \Sigma\, b_{jk} \tilde{x}_k$, where $B = (b_{jk})$ is an orthogonal matrix, such that $\tilde{x}_1 + i\tilde{x}_2$ is an analytic function on S.

Proof: In the case where the a_k are real, equation (12.12) simply says that the tangent vectors v and w are both orthogonal to the vector $a = (a_1, ..., a_n)$, where $z_k = v_k + iw_k$. By integration, the surface lies in the hyperplane orthogonal to the vector a.

In the second case, if we set $a_k = \alpha_k + i\beta_k$, the vectors α and β are real orthogonal vectors of equal length, by (12.13), and we may assume them to be unit vectors. We may then complete them to an orthonormal set of vectors which we use to form the columns of the matrix B. Then introducing the functions ϕ_k in terms of local isothermal parameters, equation (12.12) takes the form

$$0 = \Sigma \, \bar{a}_j \phi_j = \Sigma \, \bar{a}_j b_{jk} \tilde{\phi}_k = \tilde{\phi}_1 - i\tilde{\phi}_2 \, .$$

But this is just the Cauchy-Riemann equations for the function $\tilde{x}_1 + i\tilde{x}_2$, which proves the lemma. ♦

Let us note the following concerning the two cases which arise in the lemma. In the first case we can make an orthogonal change of coordinates so that S lies in a hyperplane $\tilde{x}_1 =$ constant, or equivalently $\tilde{\phi}_1 \equiv 0$. In the second case we have $\tilde{\phi}_1 = i\tilde{\phi}_2$, or $\tilde{\phi}_1^2 + \tilde{\phi}_2^2 \equiv 0$. In general, we are led to make the following definition.

DEFINITION. A minimal surface in E^n is *decomposable* if, with respect to suitable coordinates, we have

(12.14) $$\sum_{k=1}^{m} \phi_k(\zeta)^2 \equiv 0, \text{ for some } m < n.$$

Note that equation (12.14) implies that also

$$\sum_{k=m+1}^{n} \phi_k(\zeta)^2 \equiv 0 \, .$$

This means that if we write $E^n = E^m \oplus E^{n-m}$, then the projection of the surface into each of these subspaces is again a minimal surface, although one of them may degenerate into a constant mapping. In particular, one can always manufacture minimal surfaces in higher dimensions out of pairs of minimal surfaces in lower dimensions by this procedure. As a special case, if

$$x_1 + ix_2, \; x_3 + ix_4, \ldots, x_{n-1} + ix_n$$

are analytic functions of a variable ζ, where n is even, then the surface $x(\zeta)$ is a (decomposable) minimal surface in E^n. The example given at the end of section 2 is an illustration of this.

LEMMA 12.4. *For a minimal surface S in E^3, the following are equivalent:*

   a) S *is decomposable;*
   b) *the Gauss map of S is degenerate;*
   c) S *lies on a plane.*

Proof: Clearly a) is equivalent to c), because if S is decomposable, either $m = 1$ or $m = 2$, in which case either $\phi_1 \equiv 0$ or $\phi_3 \equiv 0$. If the Gauss map is degenerate, then the image of S lies in a hyperplane H of $P^2(\mathbf{C})$ and in the quadric Q_1. But one may easily verify that $H \cap Q_1$ consists of either one or two points, and since the image is connected, it must reduce to a point, and S lies in a plane by Lemma 12.2. ♦

If one now passes to higher dimensions, then no two of these conditions are equivalent, even if one replaces condition c) by the natural generalization, "S lies in a hyperplane." We have already seen that the latter condition corresponds to a special type of degeneracy or decomposability. As for the first two conditions, the example given at the end of section 5 is a surface for which $\phi_2 = -2i\phi_1$, so that the Gauss map is degenerate, but one can easily verify that it is not decomposable.

In the above discussion, as well as in condition (12.9) in Theorem 12.1, we have described the Gauss map in terms of the image lying in certain subsets of $P^{n-1}(\mathbf{C})$. For many purposes it is more natural to characterize the set of hyperplanes in $P^{n-1}(\mathbf{C})$ which intersect the image. One can easily verify, for example, that equation (12.9) is equivalent to the statement: "the image under the Gauss map does not intersect any hyperplane sufficiently near the hyper-plane $\Sigma\, a_k z_k = 0$." Thus Theorem 12.1 can be reworded: *"If S is a complete minimal surface, not a plane, then its image under the Gauss map intersects a dense set of hyperplanes."* *

For a closer study of the Gauss map from this point of view we note the following.

LEMMA 12.5. *Given a minimal surface S in E^n, the Gauss map $g(S)$, and a hyperplane H in $P^{n-1}(C)$, then either $g(S)$ lies in H, or else H intersects $g(S)$ at isolated points with fixed multiplicities.*

Proof: If H is given by $\Sigma\, a_k z_k = 0$, then near any point of S choose local isothermal parameters ζ, and set $\psi(\zeta) = \Sigma\, \bar{a}_k \phi_k(\zeta)$. Then either $\psi(\zeta) \equiv 0$, in which case $g(S)$ lies in H locally, and by analyticity, globally, or else $g(\zeta)$ has isolated zeros whose multiplicities are independent of the choice of parameter ζ. ◆

We are thus led to a study of the "meromorphic curve" $g(S)$, which is precisely the type of object treated in detail in the book of Weyl [1]. It is advantageous for the general theory, and particularly for the applications we wish to make, to introduce a Riemannian metric on $P^{n-1}(C)$. We may do this by defining the element of arclength $d\sigma$ of a differentiable curve $z(t)$, by the expression

$$(12.15) \qquad \left(\frac{d\sigma}{dt}\right)^2 = 2\, \frac{|z(t) \wedge z'(t)|^2}{|z(t)|^4} = 2\, \frac{\displaystyle\sum_{j<k} |z_j z'_k - z_k z'_j|^2}{(\Sigma |z_k|^2)^2} \ .$$

Up to a multiplicative factor, this is a standard metric for complex projective space.

Suppose now that we have a minimal surface S, and in terms of local isothermal parameters ζ, $z_k = \overline{\phi_k(\zeta)}$. Then (12.15) takes

---

*See Appendix 3, Section 4.

the form

(12.16)
$$\left(\frac{d\sigma}{dt}\right)^2 = \tilde{\lambda}^2 \left|\frac{d\zeta}{dt}\right|^2$$

where

(12.17)
$$\tilde{\lambda} = \sqrt{2} \frac{|\phi(\zeta) \wedge \phi'(\zeta)|}{|\phi(\zeta)|^2} \quad .$$

In particular, the image of a domain Δ in the ζ-plane has area

(12.18)
$$\tilde{A} = 2 \int\int_\Delta \frac{|\phi \wedge \phi'|^2}{|\phi|^4} d\xi_1 d\xi_2$$

with respect to this metric.

We next compute the Gauss curvature. The formula (9.5): $K = -\Delta \log \lambda / \lambda^2$, applied to the metric on S:

(12.19)
$$\frac{ds}{dt} = \lambda \left|\frac{d\zeta}{dt}\right| \quad ; \quad \lambda^2 = \frac{1}{2} |\phi(\zeta)|^2 \ ,$$

yields the expression

(12.20)
$$K = -\frac{4|\phi \wedge \phi'|^2}{|\phi|^6} = -\frac{\tilde{\lambda}^2}{\lambda^2} \quad .$$

In particular, for the total curvature, we obtain

$$\int\int K \, dA = \int\int_\Delta K\lambda^2 d\xi_1 d\xi_2 = -\int\int_\Delta \tilde{\lambda}^2 \, d\xi_1 d\xi_2 = -\tilde{A} \ .$$

In other words: *the total curvature of any part of the surface is the negative of the area of the image with respect to the metric (12.15).*

We thus have a precise generalization of the situation in three dimensions, represented by (9.11), and we may examine the form in which various theorems generalize. The methods of integral geometry allow one to relate the number of times the image under the Gauss map intersects various hyperplanes with the area of the image, hence with the total curvature of the surface. We do not go into the details here, but refer the interested reader to the discussion in the paper of Chern and Osserman [1].

APPENDIX 1

LIST OF THEOREMS

THEOREM 5.1. *Let* $f(x_1, x_2) = (f_3(x_1, x_2), ..., f_n(x_1, x_2))$ *be a solution of the minimal surface equation (2.8) in the whole* x_1, x_2-*plane. Then there exists a nonsingular linear transformation* $x_1 = u_1$, $x_2 = au_1 + bu_2$, $b > 0$, *such that* (u_1, u_2) *are (global) isothermal parameters for the surface* S *defined by* $x_k = f_k(x_1, x_2)$, $k = 3, ..., n$.

COROLLARY 1. *In the case* $n = 3$, *the only solution of the minimal surface equation in the whole* x_1, x_2-*plane is the trivial solution,* f *a linear function of* x_1, x_2.

COROLLARY 2. *A bounded solution of equation (2.8) in the whole plane must be constant (for arbitrary n).*

COROLLARY 3. *Let* $f(x_1, x_2)$ *be a solution of (2.8) in the whole* x_1, x_2-*plane, and let* $\tilde{S}$ *be the surface defined by* $x_k = \tilde{f}_k(u_1, u_2)$, $k = 3, ..., n$, *obtained by referring the surface* S *to the isothermal parameters given in Theorem 5.1. Then the functions*

$$\tilde{\phi}_k = \frac{\partial \tilde{f}_k}{\partial u_1} - i \frac{\partial \tilde{f}_k}{\partial u_2} \; , \qquad k = 3, ..., n \; ,$$

are analytic functions of $u_1 + iu_2$ *in the whole* u_1, u_2-*plane and satisfy the equation*

$$\sum_{k=3}^{n} \tilde{\phi}_k^2 \equiv -1 - c^2 \ . \qquad c = a - ib \ .$$

Conversely, given any complex constant $c = a - ib$ with $b > 0$, and given any entire functions $\tilde{\phi}_3, \ldots, \tilde{\phi}_n$ of $u_1 + iu_2$ satisfying the above equation, these may be used to define a solution of the minimal surface equation (2.8) in the whole x_1, x_2-plane.

THEOREM 7.1. *Let* Γ *be an arbitrary Jordan curve in* E^n. *Then there exists a simply-connected generalized minimal surface bounded by* Γ.

THEOREM 7.2. *Let* D *be a bounded convex domain in the* x_1, x_2-*plane, and let* C *be its boundary. Let* $g_k(x_1, x_2)$ *be arbitrary continuous functions on* C, $k = 3, \ldots, n$. *Then there exists a solution* $f(x_1, x_2) = (f_3(x_1, x_2), \ldots, f_n(x_1, x_2))$ *of the minimal surface equation (2.8) in* D, *such that* $f_k(x_1, x_2)$ *takes on the boundary values* $g_k(x_1, x_2)$.

THEOREM 8.1. *Let* S *be a complete regular minimal surface in* E^3. *Then either* S *is a plane, or else the normals to* S *are everywhere dense.*

THEOREM 8.2. *Let* S *be a complete regular minimal surface in* E^3. *Then either* S *is a plane, or else the set* E *omitted by the image of* S *under the Gauss map has logarithmic capacity zero.*

THEOREM 8.3. *Let* E *be an arbitrary set of* k *points on the sphere, where* $k \leq 4$. *Then there exists a complete regular minimal surface in* E^3 *whose image under the Gauss map omits precisely the set* E.

THEOREM 9.1. *Let* M *be a complete Riemannian 2-manifold whose Gauss curvature satisfies* $K \leq 0$, $\iint_M |K|\, dA < \infty$. *Then there exists a compact 2-manifold* $\tilde{M}$, *a finite number of points* $p_1, \ldots, p_k$ *on* $\tilde{M}$, *and an isometry between* M *and* $\tilde{M} - \{p_1, \ldots, p_k\}$.

THEOREM 9.2. *Let* S *be a complete minimal surface in* E^3. *Then either* $\iint_S K\, dA = -\infty$, *or else* $\iint_S K\, dA = -4\pi m$, $m = 0, 1, 2, \ldots$.

THEOREM 9.3. *Let* S *be a complete regular minimal surface in* E^3. *Then* $\iint_S K\, dA \leq 2\pi(\chi - k)$, *where* χ *is the Euler characteristic of* S, *and* k *is the number of boundary components of* S.

THEOREM 9.4. *There are only two complete regular minimal surfaces in* E^3 *whose total curvature is* -4π. *These are the catenoid and Enneper's surface.*

COROLLARY. *Enneper's surface and the catenoid are the only two complete regular minimal surfaces in* E^3 *whose Gauss map is one-to-one.*

THEOREM 9.5. *Let* S *be a regular minimal surface in* E^3. *and suppose that all paths on the surface which tend to some isolated boundary component of* S *have infinite length. Then either the normals to* S *tend to a single limit at that boundary component, or else in each neighborhood of it the normals take on all the directions except for at most a set of capacity zero.*

THEOREM 10.1. *Let* $F(x_1, x_2)$ *be a solution of the minimal surface equation (10.1) in a bounded domain* D. *Suppose that for every boundary point* (a_1, a_2) *of* D, *with the possible exception of a finite number of points, the relations*

$$\lim_{(x_1, x_2) \to (a_1, a_2)} F(x_1, x_2) \leq M, \qquad \overline{\lim_{(x_1, x_2) \to (a_1, a_2)}} F(x_1, x_2) \geq m$$

are known to hold. Then $m \leq F(x_1, x_2) \leq M$ *throughout* D.

THEOREM 10.2. *A solution of the minimal surface equation cannot have an isolated singularity.*

THEOREM 10.3. *Let* D *be a bounded domain in the plane. Necessary and sufficient that there exist a solution of the minimal surface equation in* D *taking on arbitrarily assigned continuous values on the boundary is that* D *be convex.*

THEOREM 10.4. *Let* D *be an arbitrary domain having a point of concavity* P *on its boundary. Then a solution of the minimal surface equation in* D *cannot tend to infinity at* P.

THEOREM 10.5. *Let* D *be a plane domain,* γ *an arc of the boundary of* D, L *the line segment joining the endpoints of* γ. *If* γ *and* L *bound a subdomain* D' *of* D, *then no solution of the minimal surface equation in* D *can tend to infinity at each point of* γ.

THEOREM 11.1. *Let* $x_3 = F(x_1, x_2)$ *define a minimal surface in the disk* D: $x_1^2 + x_2^2 < R^2$. *Let* P *be the point on the surface lying over the origin, let* K *be the Gauss curvature of the surface at* P, *and let* d *be the distance along the surface from* P *to the boundary. Then the inequality* $|K| \leq c/d^2 W_0^2$ *holds, where* c *is an absolute constant, and* W_0^2 *is the value at the origin of* $W^2 = 1 + (\partial F/\partial x_1)^2 + (\partial F/\partial x_2)^2$.

COROLLARY. *Under the same hypotheses, there exists an absolute constant* c_0 *such that* $|K| \leq c_0/R^2$.

THEOREM 11.2. *Let* $x_3 = F(x_1, x_2)$ *define a minimal surface in the region* $x_1^2 + x_2^2 > R^2$. *Then the gradient of* F *tends to a limit at infinity.*

THEOREM 11.3. *If the exterior Dirichlet problem has a bounded solution, then it is unique.*

THEOREM 11.4. *There exist continuous functions on the circle* $x_1^2 + x_2^2 = 1$ *which may be chosen to be arbitrarily smooth, such that no bounded solution of the minimal surface equation in the exterior of this circle takes on these values on the boundary.*

THEOREM 12.1. *Let* S *be a complete regular minimal surface in* E^n. *Then either* S *ia a plane, or else the image of* S *under the Gauss map approaches arbitrarily closely every hyperplane in* $P^{n-1}(C)$.

COROLLARY. *The normals to a complete regular minimal surface* S *in* E^n *are everywhere dense unless* S *is a plane.*

APPENDIX 2

aspects of stability of "H" and K

GENERALIZATIONS

One of the important roles played historically by the theory of minimal surfaces was that of a spur to obtain more general results. In the following pages we shall try to give a brief idea of some of the extensions of the theory which we have discussed. We shall group the results in several categories.

I. *Wider classes of surfaces in* E^3.

Relate to stability

A. *Surfaces of constant mean curvature.*

From a geometric point of view, the most natural generalization of surfaces in E^3 satisfying $H \equiv 0$ consists of surfaces satisfying $H \equiv c$. Certain properties extend to this class, whereas others are quite different according as $c = 0$, or $c \neq 0$,

Let us start with the Plateau problem. Theorem 7.1 was extended first by Heinz [2], who pointed out that in general, one would expect a solution only if the mean curvature is not too large compared to the size of the curve Γ. (See also the discussion below for the non-parametric case.) The results of Heinz were later improved on by Werner [1], who obtained, in particular, the following result: *Let Γ be a Jordan curve which lies in the unit sphere. Then for any value of c, $|c| < \frac{1}{2}$, there exists a simply-connected surface satisfying $H \equiv c$, bounded by Γ.*

Complete surfaces of constant mean curvature are studied in Klotz and Osserman [1] using methods similar to those of section 8. The following result is proved: *Let S be a complete surface of constant mean curvature. If $K \geq 0$ on S, then S is a plane,*

135

a sphere, or a right circular cylinder. If $K \leq 0$, then S is a cylin-der or a minimal surface. If one makes no restriction on the sign of the Gauss curvature K, then there exist complete surfaces of constant mean curvature which are none of the types listed.

Concerning non-parametric surfaces, we have first of all the following result, which may be considered the natural generaliza-tion of Bernstein's theorem. *Let* $x_3 = F(x_1, x_2)$ *define a surface of constant mean curvature* $H = c > 0$ *in a disk* $x_1^2 + x_2^2 < R^2$. *Then* $R \leq 1/c$, *with equality holding only if the surface is a hemi-sphere.* The first statement in the conclusion is due to Bernstein [1] and the second to Finn [5]. The theorem of Bers on removable singularities (Theorem 10.2) goes over without change (Finn [1]), and indeed so does the fact that every set of vanishing linear Haus-dorff measure is removable (Serrin [2]). Finally, there is an exis-tence theorem corresponding to Theorem 10.3. (See II B, below.)

B. *Quasiminimal surfaces.*

Considering minimal surfaces in non-parametric form from the point of view of partial differential equations, one is led to ask for more general equations whose solutions behave in a similar manner. Various classes of equations were studied by Bers [3] and Finn [2] with a view toward generalizing results such as Bernstein's theorem and Theorems 10.2 and 11.2 of Bers. The underlying idea is to replace conformal maps by quasiconformal maps at some point in the theory. In particular, Finn introduces a class of equations of "minimal surface type" whose solutions have the property that their Gauss map is quasiconformal. With varying additional restric-tions, these equations have been studied in Finn [2, 3, 4], Jenkins [1, 2], and Jenkins and Serrin [1].[*] It is shown in these papers that

---

[*]See Appendix 3, Section 6IB.

a large part of the theory of minimal surfaces which we have presented here can be carried over, and in fact some results were
first proved in this wider context before they were known for the
special case of minimal surfaces. Among the most interesting,
let us note the Harnack inequality and the convergence theorems
of Jenkins and Serrin [1].

If one considers parametric surfaces with the single requirement that the Gauss map be quasiconformal, one is led to the
class of surfaces introduced in Osserman [4] as "quasiminimal
surfaces." Certain results, for example Theorem 9.2, can be
shown to continue to hold in this generality. It would be interesting to know whether Bernstein's theorem is true for quasiminimal
surfaces, or whether some of the additional restrictions imposed
by Finn are really necessary.*

II. *Hypersurfaces in E^n.*

A. *Minimal hypersurfaces.*

From many points of view the most natural generalization of
surfaces in E^3 is to hypersurfaces in E^n. If one asks for the
condition that a hypersurface minimize volume with respect to all
hypersurfaces having the same boundary, then one is led to the
geometric condition that the mean curvature be zero, and the analytic condition that the surface in non-parametric form $x_n =
F(x_1, ..., x_{n-1})$ satisfy the equation

$$(1) \quad \sum_{i=1}^{n-1} \frac{\partial}{\partial x_i} \left(\frac{p_i}{W} \right) = 0, \quad p_i = \frac{\partial F}{\partial x_i}, \quad W = \left(1 + \sum_{i=1}^{n-1} p_i^2 \right)^{\frac{1}{2}}.$$

It has long been an open problem whether Bernstein's theorem
generalizes to this case, i.e., whether every solution of (1) de-

---

*See Appendix 3, Section 6IB.

fined for all $x_1, \ldots, x_{n-1}$ is linear. This was proved in the case $n = 4$ by de Giorgi [1], for $n = 5$ by Almgren [2], and for $n = 6, 7, 8$ by Simons [1]. Weaker forms of Bernstein's theorem have been proved for arbitrary n, where the conclusion that F is linear is obtained by imposing additional restrictions, such as W bounded (Moser [1]) or F positive (Bombieri, de Giorgi, and Miranda [1]). The latter paper also contains references to a number of other recent contributions to the theory of minimal hypersurfaces.[*]

An analogue of Theorem 7.2 is given in a recent paper of Miranda [1], while de Giorgi and Stampacchia [1] gave the following theorem on removable singularities: *Every solution of equation (1) in a domain D from which a compact subset E of $(n-2)$-dimensional Hausdorff measure zero has been removed extends to a solution in all of D.*[†]

Up till now it has not been possible to extend the results of sections 8 and 9 on parametric surfaces to the case of hypersurfaces.

Finally, Jenkins and Serrin [3] have obtained the following remarkable generalization of Theorem 10.3: *Let D be a bounded domain in E^{n-1} with C^2 boundary, and let H be the mean curvature of the boundary with respect to the inner normal. Then necessary and sufficient that there exist a solution of equation (1) in D taking on arbitrary assigned C^2 boundary values is that $H \geq 0$ everywhere on the boundary.*

---

[*]It appears that the problem of Bernstein's theorem has just been settled in the most unexpected fashion. Bombieri [1] has announced the construction of a counterexample showing that although Bernstein's theorem is true for $n \leq 8$, it fails to hold for $n > 8$. The details will appear in a paper of Bombieri, de Giorgi, and Giusti [1].

[†]See Appendix 3, Sections 6IIA and 6IIB.

B. *Other hypersurfaces.*

The result of Jenkins and Serrin just referred to has been further extended by Serrin [3] to a large class of equations which include the equation for surfaces of constant mean curvature. His
result is the following: Under the same hypotheses as before,
*necessary and sufficient that there exist a hypersurface of con-
stant mean curvature c corresponding to arbitrarily prescribed C^2
boundary values is that $H \geq c(n-1)/(n-2)$ at each point of the
boundary.*

We note also the following result, contained in a recent paper
of Chem [2]. *If $x_n = F(x_1, ..., x_{n-1})$ defines a hypersurface of
constant mean curvature, $H \equiv c$, over the whole space of $x_1, ...,
x_{n-1}$, then in fact $H \equiv 0$.*

III. *Minimal varieties in E^n.*

We come now to the general case of k-dimensional minimal varieties in E^n, where $2 \leq k \leq n-1$. Very little progress had been
made for the intermediate values of k, $2 < k < n-1$ until recently,
when there appeared the papers of Reifenberg [1] and Federer and
Fleming [1]. In both cases one gets away from parametric methods
(including the "non-parametric methods" discussed earlier) and
introduces suitable classes of objects which are handled chiefly
by measure-theoretic methods. In particular, Reifenberg [1, 2, 3]
showed that there always exists a compact set X having a given
boundary A in a certain precise sense, such that the set $X - A$,
outside of a possible subset of $(k-1)$-dimensional Hausdorff measure zero, is a k-dimensional minimal variety.

These methods have been carried further by Almgren [1] and by

Morrey [3]. The book of Morrey is an excellent general reference
to this theory, viewed in terms of differential equations and the
calculus of variations.

A major question concerning these methods is whether the possible singular sets actually occur. Simons [1] showed that in the
case of hypersurfaces in dimensions $n \leq 7$, the set $X - A$ is a real
analytic minimal variety without singularities. He also gave an example for $n = 8$ of a cone having a singularity at the origin which
provides a relative minimum in Plateau's problem. Now Bombieri
[1] has announced a series of results which prove that Simons' cone
represents an absolute minimum, and hence a counterexample to
the regularity of the solution.   Details are given in Bombieri,
de Giorgi and Giusti [1].

IV. *Minimal subvarieties of a Riemannian manifold.*

The theory of two dimensional minimal surfaces in a Riemannian manifold has been studied by various authors (for example,
Lumiste[1], Pinl [2]). Morrey [1] showed that the problem of Plateau could be solved also for the case of a Riemannian manifold.
More recently Morrey [2, 3] has extended the work of Reifenberg
from the euclidean to the Riemannian case. Almgren [1, 2] and
Frankel [1] have studied topological properties of minimal sub-
varieties.

We have noted earlier that a complex-analytic curve in C^n,
considered as a real two-dimensional surface in E^{2n}, is always a
minimal surface. This generalizes to the statement that a Kähler
submanifold of a Kähler manifold is a minimal variety. For this
and related statements we refer to the recent paper of Gray [1].

A major contribution to the theory of minimal varieties has
been made by Simons [1]. He derives an elliptic second order

equation which is satisfied by the second fundamental form of a
minimal variety in an arbitrary Riemannian manifold. There are a
number of applications, including the extension of Bernstein's
theorem mentioned above (in IIA), and a large part of the earlier
theory is unified and generalized.

Minimal submanifolds of spheres have been the object of particularly intensive research in the past two or three years. In addition to several of the above-mentioned papers we note the work
of Calabi [2] which applies complex-variable methods to 2-dimen-
sional minimal surfaces in an n-sphere, and the papers of Chern,
do Carmo, and Kobayashi [1], Hsiang [1,2], Lawson [1,2,3],
Otsuki [1], Reilly [1], Takahashi [1], and Takeuchi and Kobayashi
[1].

To all those whose work has not been mentioned, or barely
skimmed over in this brief survey, we apologize. To the novice
in this field we hope at least to have given some idea of the range
and the depth of current activity in the field of minimal surfaces.

APPENDIX 3

DEVELOPMENTS IN MINIMAL SURFACES, 1970–1985

1. *Plateau's Problem*

Many basic questions concerning Plateau's problem have been settled in the past decade. The fundamental existence theorem of Douglas,
Theorem 7.1, may now be sharpened to the form: *let Γ be an arbitrary*
Jordan curve in E^3. Then there exists a **regular** *simply-connected*
minimal surface bounded by Γ. (See Osserman [11], Gulliver [1],
Gulliver, Osserman, and Royden [1], and Alt [1].) Among the further
questions that arise are whether or not the surface obtained is actually
embedded (i.e., without self-intersections), and when the solution is
unique. If the curve Γ lies on the boundary of a convex body, then
Meeks and Yau [1] proved that the surface is indeed embedded, while,
independently, Almgren and Simon [1] and Tomi and Tromba [1] proved
related embedding results. Earlier, Gulliver and Spruck [1] proved the
result under the additional assumption that the total curvature of Γ was
at most 4π.

In all the above results we cannot properly speak of *the* surface
bounded by Γ (or *the* Douglas solution) since one does not in general
have uniqueness. An interesting example is provided by Enneper's
surface. In the standard representation of Enneper's surface (see p. 65,
above), if one denotes by Γ_r the image of the circle $|\zeta| = r$, then Nitsche
[10] has shown that for $1 < r < 1 + \epsilon$ there exist at least three distinct
solutions to Plateau's problem with boundary Γ_r. In the other direction,
Ruchert [1] proved that for $0 < r \leqslant 1$, the solution is unique. Ruchert's
proof makes use of a sharpened form of a uniqueness theorem due to

Nitsche [9]: *a real analytic Jordan curve in E^3 whose total curvature is at most 4π bounds a unique simply-connected minimal surface.* An interesting counterpart is a construction of Almgren and Thurston [1], who show that *given any ϵ and any positive integer g, there exists a Jordan curve Γ in E^3 with the property that the total curvature of Γ is at most $4\pi + \epsilon$, and any embedded minimal surface bounded by Γ must have genus at least g.* If one drops the requirement of simple connectivity, then one has extensive results due to Gulliver [2] on the regularity of Douglas' solutions of higher topological type (where the topological type is specified in advance), and the basic theorem of Hardt and Simon [1] proving the existence of a regular embedded minimal surface spanning an arbitrary given set of Jordan curves in E^3.

When we go to E^n, $n \geq 4$, although Douglas' basic existence theorem 7.1 still holds, both regularity and uniqueness tend to break down. An important example is the surface S in E^4 defined non-parametrically by $x_3 = \text{Re}\{R(x_1 + ix_2)^4\}$, $x_4 = \text{Im}\{Rx_1 + ix_2)^4\}$, $|x_1 + ix_2| \leq 1$, R a positive constant. As was noted at the end of §2, S is a minimal surface. Its boundary Γ is a Jordan curve. Federer [1] showed that S is in fact the unique oriented surface of least area bounded by Γ. Thus in this case, Douglas' solution is unique. On the other hand, Morgan [1] has shown that there exists a non-oriented minimal surface S^* bounded by Γ, and that for sufficiently large R there exists a one-parameter family of isometries of E^4 mapping Γ onto itself and taking S^* onto a one-parameter family of mutually distinct minimal surfaces of least area all bounded by Γ. Federer's Theorem also gives examples where Douglas' solution is not regular, but has branch points.

Even for non-parametric solutions, uniqueness may fail. Thus, in Theorem 7.2, the solution is unique when $n = 3$ by Lemma 10.1, whereas for $n \geq 4$ there exist arbitrarily smooth boundary values on the unit circle for which the solution is not unique (Lawson and Osserman [1]).

A paper of White [2] gives a generic regularity result: for almost every smooth closed curve in E^n, the unoriented area-minimizing surfaces that it bounds have no singularities.

Returning to E^3, Morgan [3] gave another example of a continuum of distinct minimal surfaces having the same boundary. In his example, the boundary is the union of four disjoint circles. On the other hand, in the paper referred to above, Hardt and Simon [1] show that if Γ is an arbitrary collection of sufficiently smooth Jordan curves in E^3, then Γ can bound only a finite number of oriented *area-minimizing* surfaces. Their result uses basic earlier work of Böhme and Tomi [1], as well as Tomi [1]. For further work on finiteness, we refer to the papers of Tromba [1], Morgan [2,5], Nitsche [III], Beeson [2], Koiso [1], and Böhme and Tromba [1].

Another area of major progress is that of regularity at the boundary. This was settled by a number of authors following the initial ground-breaking work of Hildebrandt [3]. For details we refer to the book of Nitsche [II] (Chapter V, §2.1) and to the paper of Hardt and Simon [1].

Finally we note that a totally different approach to regularity is given by Beeson [1], who applies variational arguments in a neighborhood of a branch point. For background on the subject of branch points on minimal surfaces, see Osserman [I] and Nitsche [II], V.2.2. For more on Plateau's problem, including a long list of open problems, see the lecture notes of Meeks [I].

2. *Stability*

There is a class of minimal surfaces that falls between that of the area-minimizing surfaces and that of the totality of all minimal surfaces: *stable* minimal surfaces. There are various notions of stability, but basically a stable surface is area-minimizing relative to nearby surfaces with the same boundary. Thus, stable minimal surfaces are precisely the ones that one would expect to obtain in physical experiments. In terms of our discussion in §3, minimality of a surface is equivalent to the vanishing of the first variation of area $A'(0)$ under all deformations of the surface leaving the boundary fixed. Thus, minimality is a necessary condition for a surface to be area-minimizing. However, if for some deformation the second variation is negative, $A''(0) < 0$, then there are nearby surfaces of smaller area, and the surface is called *unstable*.

The property of stability has assumed increasing importance in recent years. Among the results obtained are the following:

Barbosa and do Carmo [1] showed that for a surface S in E^3, if the image of S under the Gauss map has area less than 2π, then S is stable: the second variation of area is positive for all deformations fixing the boundary. This result has found important applications in Nitsche's theorems [9,III] on uniqueness and finiteness. It was later generalized to minimal surfaces in E^n by Spruck [1], replacing the Gauss map by the generalized Gauss map (12.7). Spruck's result was in turn improved by Barbosa and do Carmo [2], who showed that the assumption $\int\int|K|dA < 4\pi/3$ implied stability for simply-connected minimal surfaces in E^n.

The above results all involve sufficient conditions for stability. In the other direction, the assumption of stability often has important consequences. One example is a recent result of Schoen [1], who obtains an analog of inequality (11.1) for stable surfaces. Specifically, Schoen shows that *there is a universal constant c such that if S is a stable minimal surface in E^3, then for any point p in S the Gauss curvature K of S at p satisfies $|K(p)| \leq c/d^2$, where d is the distance from p to the*

boundary of S. Since a non-parametric minimal surface can easily be
shown to be area-minimizing with respect to its boundary and hence
stable, Schoen's result is an extension of Heinz inequality (11.7), which
in turn implies Bernstein's Theorem. It also implies the earlier result of
Fischer-Colbrie and Schoen [1] and do Carmo and Peng [1] that, *if S is a
complete globally stable minimal surface in E^3, then S is a plane.* The
assumption "*S* globally stable" means that every relatively compact
domain *D* on *S* is stable.

Another class of surfaces that are area-minimizing, and therefore
globally stable, is complex holomorphic curves in $\mathbb{C}^n$, considered as real
surfaces in E^{2n} (Federer [1]). A paper of Micallef [1] makes a number of
important contributions to the study of stability, including the proof
that a complete globally stable minimal surface in E^4 must be a holomorphic complex curve with respect to some orthogonal complex structure on E^4 (which is thereby identified with $\mathbb{C}^2$), provided that some
further condition is satisfied, such as a restriction on the Gauss map,
finite total curvature, or that the surface is non-parametric. This last
condition was obtained independently by Kawai [1]. It shows in particular that the surfaces, such as (5.19), obtained as global solutions of
the minimal surface equation must be unstable, even though they are
non-parametric.

The paper of Schoen [1] applies to minimal surfaces not only in E^3
but also in a three-dimensional Riemannian manifold. The study of
stable minimal surfaces in Riemannian manifolds has been a very
fruitful one in recent years and has led to some of the most important
advances. These will be discussed below in Sections 6.IV.A and 6.V.

3. *Isoperimetric inequalities*

There are two roles that isoperimetric inequalities have played in
recent work on minimal surfaces. First, isoperimetric inequalities for

domains on minimal surfaces have been applied in a number of ways. (For a discussion of such applications, see §4 of Osserman [III].) Second, isoperimetric inequalities on arbitrary surfaces have been applied to the theory of minimal surfaces. For example, the stability results of Barbosa and do Carmo [1, 2, 3] and Spruck [1] depend on the use of iso-perimetric inequalities on the image of a minimal surface under the (generalized) Gauss map.

It has been a long-standing conjecture that the classical iso-perimetric inequality $L^2 \geq 4\pi A$ should hold for an arbitrary domain on a minimal surface in E^n, where A is the area of the domain, and L the length of the boundary. Until recently that was only known for an oriented domain bounded by a single Jordan curve. (New and simpler proofs of that have been given by Chakerian [1] and Chavel [1].) Then it was proved for doubly-connected domains, first in E^3 (Osserman and Schiffer [1]) and then in general (Feinberg [1]). That in turn implied the result for minimal Möbius strips (Osserman [13]). Except under further hypotheses, the sharp isoperimetric inequality for domains of arbitrary topological type on minimal surfaces is still not known. (See Li, Schoen, and Yau [1] for certain hypotheses that suffice.) For more details, see Nitsche [II], §323, and Osserman [III], §4.

4. Complete minimal surfaces

One of the most striking results of recent years is a theorem of Xavier [1] that goes a long way toward answering the questions on p. 74, above, concerning the gap between Theorems 8.2 and 8.3. What Xavier shows is that *the normals to a complete minimal surface in E^3, not a plane, can omit at most six directions.* The method of proof is quite different from that used in the proof of Theorem 8.2, and it does not seem amenable to reducing the number of omitted values to four, which by Theorem 8.3 would be best possible value. Thus, it still remains an open question to determine the precise number of values that the normals to a complete non-planar minimal surface in E^3 can omit.

Xavier's Theorem represents a very strong generalization of Bernstein's Theorem (Corollary 1 of Theorem 5.1). Two other generalizations of Bernstein's Theorem in different directions have been proved recently. One is the theorem on complete globally stable surfaces discussed above. The other is a theorem of Schoen and Simon [2]. Instead of a condition on the Gauss map or on stability, an assumption is made on area growth. Specifically, they prove the following: *let S be a simply-connected complete minimal surface embedded in E^3. For some fixed point p in S, let $S_p(r)$ denote the component containing p of the intersection of S with a ball of radius r centered at p. Let A(r) be the area of $S_p(r)$. If $A(r) < Mr^2$ for some fixed M, then S must be a plane.* Using the area-minimizing property of non-parametric minimal surfaces, and comparing $S_p(r)$ with the area of a domain on the sphere of radius r having the same boundary, one sees easily that the classical Bernstein Theorem is a consequence of the Schoen-Simon Theorem.

An exciting recent development has been the discovery of the first new complete *embedded* minimal surface of finite genus in over two hundred years. The previously known examples were the plane, the catenoid, and the helicoid. A paper of Jorge and Meeks [1] studied necessary conditions on the Weierstrass representation for the resulting surface to be a complete embedded minimal surface. Using those conditions, Costa [1] showed that there exists a constant c such that the choice of $f = \mathbf{p}$, $g = c/\mathbf{p}'$ in the Weierstrass representation (8.2), (8.6), where $\mathbf{p}$ is the elliptic $\mathbf{p}$-function of Weierstrass based on the unit square, gives a complete minimal surface of genus one in E^3, with three embedded ends. Hoffman used computer graphics to obtain excellent pictures of Costa's surface, showing clearly that it was embedded. (See Peterson [1].) Hoffman and Meeks [1] gave an analytic proof of the embeddedness, and subsequently [2] found complete embedded minimal surfaces of every genus.

Returning to the distribution of normals, we note that Gackstatter [1] showed that *the normals to a complete abelian minimal surface in E^3, not a plane, can omit at most four directions.* Complete abelian minimal surfaces are a generalization of complete surfaces of finite total curvature for which the functions φ_k of (4.6) extend to be meromorphic on a compact Riemann surface, but the $x_k = \text{Re} \int \varphi_k$ need not be single-valued. Since the surfaces of Theorem 8.3 are abelian, the number "4" of Gackstatter's theorem is sharp.

Finally, we refer to the papers of Meeks [3], Jorge and Meeks [1], Gulliver [3], and Gulliver and Lawson [1] for further results on complete minimal surfaces of finite total curvature in E^3.

Complete minimal surfaces in E^n have been studied by various authors. Among recent results, we note the following.

Gackstatter [2] studied relationships between various quantities associated with a complete minimal surface S of finite total curvature in E^n. Let m be the dimension of the smallest affine subspace of E^n containing S. Let k be the number of boundary points of S, and g its genus. Then the total curvature of S satisfies

$$\iint_S K dA \leqslant (3 - m - k - 4g)\pi.$$

This result complements inequality (9.22) in Theorem 9.3, which holds in E^n as well as E^3 (Chern and Osserman [1]). Combining the two, one can prove the following result, complementing Theorem 9.4: *if $\iint_S K dA = -4\pi$, then either S is simply-connected and $m \leqslant 6$, or else S is doubly-connected ($g = 0$, $k = 2$) and $m \leqslant 5$.* This theorem was first proved by C. C. Chen [3] using other arguments. One can in fact give a complete characterization of the surfaces with total curvature -4π (Hoffman and Osserman [1]). The doubly-connected ones are particularly interesting. They are a kind of generalized catenoid, in the sense that they are all generated by a one-parameter family of ellipses (or circles) that are obtained by intersecting the surface with the set of all hyperplanes parallel to a given fixed one. In that connection we note another result of C. C. Chen [2]: *a minimal surface S in E^n that is isometric to a catenoid is in fact itself a catenoid lying in a three-dimensional affine subspace.*

Finally we note that the complete minimal surfaces of total curvature -2π have a very simple characterization: *let (z,w) be coordinates in $\mathbb{C}^2$. For each real $c > 0$, the graph of the function $w = cz^2$ represents a minimal surface in E^4 with total curvature -2π. Every complete minimal surface in E^n with total curvature -2π lies in a four-dimensional affine subspace and is congruent to one of the above surfaces.* This theorem is a somewhat sharpened form of a theorem of C. C. Chen [3]. For this and further related results we refer to Hoffman and Osserman [1].

The above results characterizing complete minimal surfaces with total curvature -2π and -4π follow from a general characterization of complete minimal surfaces of genus zero and given total curvature (Hoffman and Osserman [1], Proposition 6.5). In particular, one has a general construction for genus zero minimal surfaces. As mentioned on p. 89, above, it would be interesting to have more examples in the higher genus case. A paper of Gackstatter and Kunert [1] shows that *given any compact Riemann surface $\overline{M}$, there exist points $p_1, \ldots p_k$, and a complete minimal surface S of finite curvature in E^3 defined by a map $x(p): M \rightarrow E^3$, where $M = \overline{M} - \{p_1, \ldots p_k\}$.* In fact, their method gives many such surfaces for any given $\overline{M}$, including a continuum of conformally distinct types. What still remains to be studied is to what degree one can prescribe the points $p_1, \ldots, p_k$. (See also Chen and Gackstatter [1], and, for non-orientable surfaces, Oliveira [1].)

With no assumption on finite total curvature, we have very recent extensions of Xavier's Theorem to E^4 by Chen [4] and to E^n by Fujimoto [1]. Fujimoto shows that *if S is a complete minimal surface in E^n with non-degenerate Gauss map g, then g(S) cannot fail to intersect more than n^2 hyperplanes in general position.* In a second paper, Fujimoto [2] gives a remarkable generalization of all these Picard-type theorems in the form of Nevanlinna-type theorems for the Gauss map.

5. *Minimal graphs*

Many results on minimal graphs in E^3 extend only partially, or not at all, to higher codimension.

Theorem 7.2 asserts the existence of a solution to the minimal surface equation in a convex plane domain corresponding to an arbitrary set of continuous boundary functions. For the case of a single boundary function, it follows from Lemma 10.1 that the solution is unique. It turns out that in the general case uniqueness does not hold. In fact, even when D is the unit disk, one can show that *there exists a pair of real analytic functions on the boundary of D to which correspond three distinct solutions in D of the minimal-surface equation* (Lawson and Osserman [1]).

Another result that fails when going from E^3 to E^n is Bers' Theorem (Theorem 10.2) asserting that a solution of the minimal-surface equation cannot have an isolated singularity. There are simple counterexamples as soon as $n = 4$, such as the complex function $w = 1/z$, considered as a pair of functions of two real variables. However, one does have the following result (Osserman [12]): *let f(x₁,x₂) be a vector solution of the minimal surface equation (2.8) in* $0 < x_1^2 + x_2^2 < \epsilon$ *for some* $\epsilon > 0$*. Suppose that all components of f with at most one exception extend continuously to the origin. Then f extends to the origin, is smooth there, and satisfies (2.8).* This result contains as special cases both Bers' Theorem and the following result proved independently by Harvey and Lawson [1]: *if f(x₁,x₂) is continuous in* $x_1^2 + x_2^2 < \epsilon^2$ *and is a solution of (2.8) in* $0 < x_1^2 + x_2^2 < \epsilon^2$, *then f is a solution in the full disk.*

Finally we note that Simon [2] has generalized the removable singularity result of Nitsche and of de Giorgi and Stampacchia (see p. 98, above) by eliminating the hypothesis that the exceptional set E be a compact subset of D.

6. *Generalizations*

We follow here the order adopted in Appendix 2, and list just a few of the subsequent results most pertinent here.

I. *Wider classes of surfaces in* E^3

A. *Surfaces of constant mean curvature*

A whole new approach to Plateau's problem for surfaces of constant mean curvature was devised by Wente [1] and elaborated in later papers of Steffen [1,2] and Wente [2]. (See the last of these for details and further references.) Wente's method involves minimizing area subject to a volume constraint, in contrast to the earlier method of Heinz [2] based on a variational problem with the mean curvature H prescribed in advance.

An analog of Theorem 8.1 has been proved by Hoffman, Osserman, and Schoen [1]. They show that *if* S *is a complete surface of constant mean curvature in* E^3 *whose Gauss map lies in a closed hemisphere, then* S *is either a plane or a right circular cylinder.* Examples of complete surfaces of revolution of constant mean curvature show that the Gauss map can lie in an arbitrarily narrow band about an equator.

An important observation due to Ruh [1] is that a surface has constant mean curvature if and only if its Gauss map is a *harmonic map.* For details on this and the general theory of harmonic maps see Eells and Lemaire [1,2]. (See also the comments on harmonic maps in VI below.)

Kenmotsu [1] derived a representation theorem for surfaces of constant mean curvature, similar to the Weierstrass representation theorem of Lemmas 8.1 and 8.2. He proved that *if g is a harmonic map of a simply-connected plane domain D into the unit sphere* Σ*, then there exists a surface S of constant mean curvature in* E^3 *defined as a branched immersion x: D* $\rightarrow$ E^3 *where the coordinates in D are iso-thermal parameters on S and the map g is the composition of the immersion map D* $\rightarrow$ *S with the Gauss map S* $\rightarrow$ Σ. More generally, for arbitrary surfaces in E^3 of variable mean curvature H, Kenmotsu derives an integrability condition relating H with the Gauss map, and obtains a generalized Weierstrass representation theorem. For further results along these lines, see Hoffman and Osserman [4].

Although somewhat further afield, the most striking recent result on surfaces of constant mean curvature is the answer by Wente [3] to an old problem of Heinz Hopf: Are there any compact immersed surfaces of constant mean curvature in E^3 other than the standard sphere? Wente showed that there is such a surface in the form of an immersed torus.

Several notions of stability are possible for surfaces of constant mean curvature. For a discussion and some recent results, see Barbosa and do Carmo [5], Palmer [1], and da Silveira [1].

B. *Quasiminimal surfaces*

The question raised on p. 137, above, has been settled by Simon

([4], Theorem 4.1) who showed that Bernstein's Theorem holds for arbitrary quasiminimal surfaces. The paper of Schoen and Simon [2] referred to earlier, where Bernstein's Theorem is generalized by imposing a bound on area growth rather than a non-parametric representation, is actually valid for all quasiminimal surfaces. Another paper of Simon [5] gives generalizations of Heinz' inequality (11.7) and of Bers' Theorems, Theorem 10.2 and 11.2, for quasiminimal surfaces that are defined via solutions to equations of "mean curvature type." This class of equations is considerably broader than similar ones considered earlier by Finn [2,4] and Jenkins and Serrin [1,2].

C. *Complete surfaces of finite total curvature*

A recent paper of White [3] shows that many of the results of Chapter 9 concerning the Gauss map and total curvature of a complete surface hold in great generality. Without assuming minimality or any other local condition, White gives new proofs and generalizations of Theorems 9.1 and 9.2, and Lemma 9.5. He assumes only that S is a complete surface in E^3 satsfying the condition $\int_S (2H^2 - K)dA < \infty$. He also obtains analogous results in E^n, generalizing those of Chern and Osserman [1].

A paper of Osserman [14] shows that a surface in E^n given by a polynomial map must have finite total curvature; if it is in addition a regular complete surface, then it must be conformally the plane.

II. *Hypersurfaces in E^n*

A. *Minimal hypersurfaces*

One of the major advances of the past decade was a paper of Schoen, Simon, and Yau [1] in which they obtain pointwise curvature estimates generalizing Heinz' inequality (11.7) to non-parametric minimal hypersurfaces in E^n for $n < 6$. An immediate consequence was a new proof of Bernstein's Theorem in those dimensions. But more important were the methods used, which have led to many further results (see IV.A, below). A number of improvements on the original paper were given by Simon [1,3].

A higher-dimensional version of the parametric Bernstein Theorem, Theorem 8.1, has been given by Solomon [1]. In fact, he gives a finite version, analogous to the parametric version of Theorem (11.1) (Osserman [2]) in the following form: *let M be a smooth area-minimizing hypersurface of E^n. Suppose that the first cohomology class of M is zero, and that the Gauss map of M omits a neighborhood U of some great $(n-3)$-dimensional sphere on the $(n-1)$-sphere. Then there exists a constant c depending on n and U such that for any point p of M, if d is the distance from p the boundary of M and if $\kappa_1, \ldots, \kappa_{n-1}$ are the principal curvatures of M at p, then*

$$\kappa_1^2 + \cdots + \kappa_{n-1}^2 \leq c/d^2.$$

A paper of Morgan [5] gives a number of finiteness theorems for the set of solutions to Plateau's problem for various classes of minimal hypersurfaces in E^n, for $n \leq 7$.

In another direction, the paper of Simon [2] sharpening the removable singularity theorem of de Giorgi and Stampacchia holds in all dimensions.

A question that has been studied by a number of authors is the following: given a Riemannian metric, when can it be realized on a minimal hypersurface in E^n? The case $n = 3$ was treated by Ricci-Curbastro [1], while higher dimensions were considered by Pinl and Ziller [1], and Barbosa and do Carmo [4]. Further conditions, as well as a review of the whole subject, are contained in Chern and Osserman [2].

B. *Other hypersurfaces*

In the papers of Simon [2,3] referred to above, he shows that the generalized Heinz inequality, Bernstein's Theorem, and the de Giorgi–Stampacchia Theorem all hold for broader classes of hypersurfaces. In particular, he extends the results of Jenkins [1] for parametric elliptic functionals in two dimensions to three dimensions, and he shows that a Heinz inequality and the Bernstein Theorem hold through dimension 7 for the non-parametric Euler-Lagrange equation of any integrand whose associated parametric functional is close enough to the area integrand.

A higher-dimensional version of Hopf's question referred to earlier (6.I.A) was answered by W.-Y. Hsiang [3], who showed that for all $n \geqslant 3$ there exist non-standard immersions of the n-sphere in E^{n+1} with constant mean curvature. (Hopf had shown that to be impossible when $n = 2$.)

III. *Minimal varieties in E^n*

A major breakthrough was the paper of White [1], solving the classical Plateau problem in higher dimensions. Specifically, any smooth map of the $(n-2)$-sphere into E^n, for $4 \leqslant n \leqslant 7$, extends to a Lipschitz map of the $(n-1)$-ball that minimizes $(n-1)$-dimensional area among all such maps. In fact, in all dimensions $n \geqslant 4$, the infimum of areas obtained from Lipschitz maps (the parametric problem) is the same as one gets using integral currents or singular chains. This result contrasts strongly with the situation for E^3, where, for example, a higher-genus surface spanning a given Jordan curve may have much less area than the parametric (Douglas) solution, obtained by mapping a disk.

Some of the most interesting other results have concerned minimal graphs of arbitrary dimension and codimension. An m-dimensional graph in E^n is given by a set of functions $x_k = f_k(x_1, \ldots, x_m)$, $k = m+1$, ..., n, or by a vector function $f(x)$, where $x = (x_1, \ldots, x_m)$, $f = (f_{m+1}, \ldots, f_n)$. The corresponding submanifold is minimal in E^n if and only if f satisfies the minimal surface equation

(*)
$$\sum_{i,j=1}^{m} g^{ij} \frac{\partial^2 f}{\partial x_i \partial x_j} = 0,$$

where $(g^{ij}) = (g_{ij})^{-1}$ and $g_{ij} = \delta_{ij} + \frac{\partial f}{\partial x_i} \cdot \frac{\partial f}{\partial x_j}$. The equation (2.8) is the special case $m = 2$ of this equation. (See Osserman [9], p. 1099, and [V].)

One of the surprising results for minimal varieties of higher codimension concerns the Dirichlet problem. Lawson and Osserman [1] showed that one can prescribe a set of three quadratic polynomials on the boundary of the unit ball in E^4 which are not the boundary values of any solution to the minimal surface system over the interior of the ball. Thus, Theorem 7.2, which holds for dimension 2 and arbitrary codimension, as well as for arbitrary dimension and codimension 1, fails for dimension 4 and codimension 3. Another result in the same paper is that there is a (specifically given) minimal cone of dimension 4 and codimension 3, which represents a solution of the minimal-surface system everywhere except at the origin, where it is continuous but has a non-removable singularity.

A Bernstein-type theorem in arbitrary dimension and codimension was proved by Hildebrandt, Jost, and Widman [1]: *Let $f(x)$ be a solution of the minimal surface equation (*) over all of E^m. Suppose there exists a number $\beta_0 < 1/\cos^p(\pi/2\sqrt{\kappa p})$, where $p = \min\{m, n-m\}$ and $\kappa = 1$ if $n = m+1$, $\kappa = 2$ if $n > m+1$. If $[\det(g_{ij})]^{1/2} \leq \beta_0$ everywhere, then each component f_j of f is a linear function of $x_1, \ldots, x_m$.*

Thus, the only entire solutions come from affine subspaces, provided a suitable gradient bound holds. In the hypersurface case $n = m + 1$, the condition reduces to a uniform gradient bound on the defining function, and the theorem reduces to that of Moser [1].

One problem in dealing with minimal submanifolds of high dimension and codimension is the paucity of examples. In that respect, recent work of Harvey and Lawson [1,3] is of special interest. They show that a certain class of closed differential forms can be used to single out submanifolds of euclidean spaces that are area-minimizing in their homology class. A special case of their construction is the family of Kähler submanifolds of $\mathbb{C}^n$, where $\mathbb{C}^n$ is identified with E^{2n}. They give other explicit examples, along with the partial differential equations that must be satisfied by non-parametric submanifolds in each class. As a special case they recover the minimal cone of Lawson and Osserman referred to above, which is thereby not only a solution of the minimal surface equation, but absolutely area-minimizing with respect to its boundary.

For absolutely area-minimizing submanifolds (and more generally, integral currents), Almgren [4] has recently proved the long-sought sharp isoperimetric inequality, with the same constant as obtained for an open subset of euclidean space of the same dimension.

Concerning complete minimal submanifolds, there are some striking recent results of Anderson [3]. He shows that the main theorems concerning the structure and the Gauss map for complete minimal surfaces of finite total curvature, proved in Chapter 9 for surfaces in E^3 and extended by Chern and Osserman [1] to surfaces in E^n, have extensions to minimal submanifolds of arbitrary dimension and codimension.

IV. *Minimal subvarieties of a Riemannian manifold*

The move from euclidean to more general Riemannian spaces as the ambient manifold represents undoubtedly the area of greatest activity in recent years. One of the main differences is that it allows minimal submanifolds to be compact. From the vast amount of work that has been done, we select just a few results.

A. *Existence theorems*

Using a combination of the methods of geometric measure theory, in particular as developed by Almgren [1], and those of Schoen, Simon, and Yau [1], Pitts [I] has obtained the most striking general existence theorems, including the following: *let M be a compact n-dimensional Riemannian manifold of class C^k, where $3 \leqslant n \leqslant 6$ and $5 \leqslant k \leqslant \infty$. Then there exists a non-empty compact embedded minimal hypersurface of class C^{k-1} in M.* Using a somewhat different approach, Schoen and Simon [1] were able to extend Pitts' results to $n \leqslant 7$, as well as to prove an important regularity theorem for stable minimal hypersurfaces in arbitrary dimension. In both papers stability plays a fundamental role, as it does in the paper of Schoen, Simon, and Yau [1], much of which applies to stable minimal hypersurfaces in an arbitrary Riemannian manifold.

There are a number of more special but very important existence theorems. Among them we note the theorem of Lawson [6], that there exist compact minimal surfaces of every genus in the 3-sphere, and a kind of dual result of Sacks and Uhlenbeck [1], proving the existence of generalized minimal surfaces of the type of the 2-sphere in a broad class of Riemannian manifolds. A later paper of Sacks and Uhlenbeck [2] proves existence of minimal immersions of higher-genus compact surfaces. Results of a similar nature were also proved by Schoen and Yau

[1], and applied to the study of three-dimensional manifolds (see V.A.B, below).

B. *Minimal surfaces in constant curvature manifolds*

Minimal surfaces in spheres continue to be studied extensively. For recent results and references to earlier ones, see the papers of Barbosa [1] and Fischer-Colbrie [1]. There has also been some work on minimal surfaces in hyperbolic space and in compact flat manifolds. For the latter, see Meeks [1,2], Micallef [2], and Nagano and Smyth [1], and further references there.

Among the topics covered, we may mention:

a) an analog of Theorems 8.1 and 8.2 stating that a compact minimal submanifold in the sphere must be a lower-dimensional great sphere if the normals omit a large enough set. The first theorem of that type is in Simons [1]; for the best results to date and for earlier references see Fischer-Colbrie [1].

b) stability: again the first results are due to Simons [1], and then later, Lawson and Simons [1]. We note in particular the fact that *there does not exist any stable compact minimal submanifold on a standard sphere.* Extensions of the Barbosa–do Carmo Theorem [1] have been made by a number of authors, in particular Barbosa and do Carmo [2,3], Mori [1], and Hoffman and Osserman [2]. For a more detailed survey of stability, see do Carmo [I]. In V, below, we give a number of applications of stability results.

c) the "spherical Bernstein problem": Hsiang [4] has shown that for $n = 4,5,6$, there exist minimal hyperspheres embedded in S^n, different from the great hyperspheres.

C. *Foliations with minimal leaves*

Minimal surfaces have turned up in a somewhat surprising manner as leaves of a foliation. In particular, a number of recent papers treat the question of characterizing those foliations such that there exists a Riemannian metric for which the leaves of the given foliation are all minimal submanifolds (see Rummler [1], Sullivan [1], Haeflinger [1], and Harvey and Lawson [4]).

V. *Applications of minimal surfaces*

In recent years the theory of minimal surfaces has been applied to the solution of a number of important problems in other parts of mathematics. We give here several examples.

A. *Topology*

In a series of papers, Meeks and Yau [1–5] have shown how adept use of solutions to the Plateau problem in Riemannian manifolds can lead to important consequences of a purely topological nature. The most striking example was their part in a string of results which when combined led to the solution of a long-standing problem in topology known as the Smith Conjecture (see Meeks and Yau [5]). Similar methods are used to obtain other purely topological results in the theory of 3-manifolds in a recent paper of Meeks, Simon, and Yau [1].

In a somewhat different direction, Schoen and Yau [1,2] use the existence of certain area-minimizing surfaces to obtain topological obstructions to the existence of metrics with positive scalar curvature on a given manifold. The stability of the minimizing surface plays a key role in the argument. More recently, Gromov and Lawson [1] have made a somewhat different use of stable minimal surfaces to study existence of metrics with various conditions on the scalar curvature. Those results are in turn used to derive topological properties of stable minimal hypersurfaces in manifolds with lower bounds on the scalar curvature. The existence of stable minimal 2-spheres is used to derive a homotopy result by J. D. Moore [1]. Other results of a related nature are due to Lawson and Simons [1] and Aminov [1]. Still other applications of minimal surfaces to topology have been given by Hass [1,3], and Nakauchi [1].

B. *Relativity*

By much more delicate arguments, but fundamentally an extension of those used in the papers referred to above, Schoen and Yau [3,4] were able to prove a well-known conjecture in general relativity, the "positive mass conjecture." Using results in these papers, they later obtained a mathematical proof of the existence of a black hole (Schoen and Yau [5]).

Other applications to relativity are given by Frankel and Galloway [1].

C. *Geometric inequalities*

Let C be a Jordan curve in E^n and let B be a closed set such that B and C are linked. (When $n = 3$, B would typically be another closed curve.) Gehring posed the problem of showing that if the distance between B and C is r, then the length L of C satisfies $L \geq 2\pi r$. Several proofs of this inequality were given, including one (Osserman [13]) that used the solution of Plateau's problem for C and the isoperimetric inequality on the resulting minimal surface. It was pointed out in that paper that the same argument would yield an analogous result in all dimensions, provided one has a parametric solution to Plateau's problem and a sharp isoperimetric inequality for the resulting surface. Neither of those results was available at the time, but they have since been proved by White [1] and Almgren [4], respectively. In the meantime a somewhat different proof of Gehring's inequality was obtained by Bombieri and Simon [1], also using minimal surfaces, and a strengthening of the result was given by Gage [1]. A generalization of Gehring's inequality was subsequently obtained by Gromov [1], p. 106, as part of a major new approach to whole classes of geometric inequalities. Solutions to generalized Plateau problems are basic to Gromov's arguments.

VI. *Harmonic maps*

The class of harmonic mappings has proved to be of increasing importance in recent years. Harmonic maps have many ties to minimal surfaces. First of all they represent a direct generalization, in that, *if a map f: M → N is an isometric immersion of one Riemannian manifold into another, then f is harmonic if and only if f(M) is a minimal submanifold of N.* When M is two-dimensional, the same result holds if f is assumed to be conformal, rather than an isometry. Slightly more generally, one has the following (Hoffman and Osserman [2]): *let f: M → N be a conformal map where the conformal factor ρ is a smooth non-negative function with ρ > 0 except on a set of measure zero. Then if* dim $M = 2$, *f is harmonic if and only if f(M) is a generalized minimal submanifold of N; i.e., f is an immersion almost everywhere with mean curvature zero; if* dim $M > 2$, *then f is harmonic if and only if f is homothetic (ρ is constant on each connected component of M) and f(M) is a minimal submanifold of N.*

Given a map $f: M → N$, we may assume that N is embedded isometrically in some E^n. If dim $M = 2$, we may choose local isothermal parameters u_1, u_2 on M, and thus get a representation of f in the form $x(u_1, u_2)$, where $x = (x_1, \ldots, x_n)$. We may form the functions $\varphi_k(\zeta) = (\partial x_k/\partial u_1) - i(\partial x_k/\partial u_2)$, as in (4.6), and we define $\varphi(\zeta) = \sum_{k=1}^{n} \varphi_k^2(\zeta)$. It turns out that *if f is a harmonic map, then φ is a holomorphic function.* (See Chern and Goldberg [1], §5, Sacks and Uhlenbeck [1], Prop. 1.5, and T. K. Milnor [1].) Furthermore, under change of isothermal parameters, φ behaves like the coefficient of a quadratic differential

$\varphi(\zeta)d\zeta^2$. As a corollary, if M is the standard 2-sphere S^2, it follows that $\varphi(\zeta) \equiv 0$. But that means precisely that the map f is conformal. Since the image of a conformal harmonic map is a minimal surface, it follows that *the image of any harmonic map $f: S^2 \to N$ is a minimal surface in N* (Chern and Goldberg [1], Prop. 5.1).

Another link between harmonic maps and minimal surfaces is the fact that *a map $f: M \to N$ is harmonic if and only if the graph of f is minimal in $M \times N$* (Eells [1]).

We note also that if a foliation is defined by a Riemannian submersion $f: M \to N$, then f is harmonic if and only if the leaves are minimal in M. (Eells and Sampson [1]. For related results, see Kamber and Tondeur [1].)

Finally, we note the basic theorem of Ruh and Vilms [1]: *let M be a submanifold of E^n. Then the generalized Gauss map g of M into the Grassmannian is a harmonic map if and only if M has parallel mean curvature. In particular, g is harmonic if M is minimal.*

For further basic bacts and references on harmonic maps, see Eells and Sampson [1] and Eells and Lemaire [1].

Among the important recent applications of harmonic maps that have been made, we mention:

A. Hildebrandt, Jost, and Widman [1] proved a Liouville-type theorem for harmonic maps, and by applying it to the Gauss map via Ruh and Vilms, they obtained the Bernstein-type theorem for arbitrary dimension and codimension cited in III above.

B. Sacks and Uhlenbeck [1,2] have proved existence theorems for harmonic maps together with arguments concerning conformal structure to get a conformal harmonic map which is thereby a minimal surface.

C. Harmonic maps have recently been studied by physicists (see

Misner [1] for a discussion of their relevance as models for physical theories). In particular, a number of physicists have studied the question of characterizing all harmonic maps of the standard 2-sphere S^2 into complex projective space $\mathbb{C}P^n$ (Din and Zakrzewski [1,2], Glaser and Stora [1]). By virtue of the result mentioned above, that all such maps are conformal, the question is equivalent to that of finding all minimal 2-spheres in $\mathbb{C}P^n$. Inspired in part by the work of the physicists, Eells and Wood [1] gave a complete classification of harmonic maps of a class of compact Riemann surfaces, including the sphere, into $\mathbb{C}P^n$. A different approach to the same problem was given by Chern and Wolfson [1].

REFERENCES

L. V. Ahlfors and L. Sario

1. *Riemann Surfaces*, Princeton University Press, Princeton, New Jersey, 1960

F. J. Almgren, Jr.

1. The theory of varifolds; a variational calculus in the large for the k-dimensional area integrand (preprint).

2. Some interior regularity theorems for minimal surfaces and an extension of Bernstein's theorem, *Ann. of Math.*, *84* (1966), 277-292.

3. *Plateau's Problem*, W. A. Benjamin, Inc., New York, 1966.

E. F. Beckenbach

1. Minimal surfaces in euclidean n-space, *Amer. J. Math.*, *55* (1933), 458-468.

E. F. Beckenbach and G. A. Hutchison,

1. Meromorphic minimal surfaces, *Pacific J. Math.*, *28* (1969), pp. 17-47.

S. Bernstein

1. Sur les surfaces définies au moyen de leur courbure moyenne ou totale, *Ann. Sci. l'Ecole Norm. Sup.*, *27* (1910), 233-256.

2. Sur les équations du calcul des variations, *Ann. Sci. Ecole Norm. Sup. (3) 29* (1912), 431-485.

3. Sur un théorème de géométrie et ses applications aux équations aux dérivées partielles du type elliptique, *Comm. de la Soc. Math. de Kharkov* (2éme sér.) *15* (1915-1917), 38-45. (Also German translation: Uber ein geometrisches Theorem und seine Anwendung auf die partiellen Differentialglei-

chungen vom elliptischen Typus, *Math. Z.*, *26* (1927), 551-558.)

L. Bers

1. Isolated singularities of minimal surfaces, *Ann. of Math.*, *(2) 53* (1951), 364-386.

2. Abelian minimal surfaces, *J. Analyse Math.*, *1* (1951), 43-58.

3. Non-linear elliptic equations without non-linear entire solutions, *J. Rational Mech. Anal.*, *3* (1954), 767-787.

L. Bianchi

1. *Lezioni di geometria differenziale II*, Pisa 1903.
   (Also German translation: *Vorlesungen über Differentialgeometrie*, 2nd ed., Teubner Verlag, Leipzig, 1910.)

E. Bombieri

1. *Lecture at Séminaire Bourbaki*, February 1969.

E. Bombieri, E. de Giorgi, and M. Miranda

1. Una maggiorazione a priori relativa alle ipersurfici minimali non parametriche, *Arch. Rational Mech. Anal.*, 32 (1969), 255-267.

E. Bombieri, E. de Giorgi, and E. Giusti

1. Minimal cones and the Bernstein problem, *Invent. Math.*, 7 (1969), 243-268.

E. Calabi

1. Quelques applications de l'analyse complexe aux surfaces d'aire minima (together with *Topics in Complex Manifolds*" by H. Rossi) Les Presses de l'Univ. de Montréal, 1968.

2. Minimal immersions of surfaces in euclidean spheres, *J. Diff. Geom. 1* (1967), 111-125.

Y.-W. Chen

1. Branch points, poles and planar points of minimal surfaces in R^3, *Ann. of Math.*, 49 (1948), 790-806.

S.-S. Chern

1. Minimal surfaces in an euclidean space of N dimensions, pp. 187-198 of *Differential and Combinatorial Topology*, A Symposium in Honor of Marston Morse, Princeton University Press, Princeton, N.J., 1965.

2. On the curvatures of a piece of hypersurface in euclidean space, *Abh. Math. Seminar*, Univ. Hamburg, *29* (1965), 77-91.

S.-S. Chern, M. do Carmo, and S. Kobayashi

1. Minimal submanifolds of a sphere with second fundamental form of constant length, pp. 59-75 of *Functional Analysis and Related Fields* (Proc. Conf. for Marshall Stone), Springer, New York, 1970.

S.-S. Chern, and R. Osserman

1. Complete minimal surfaces in euclidean n-space, *J. d'Analyse Math.*, *19* (1967), 15-34.

R. Courant

1. Plateau's problem and Dirichlet's Principle, *Ann. of Math.*, *38* (1937), 679-725.

2. *Dirichlet's principle, conformal mapping, and minimal surfaces*, Interscience, New York, 1950.

R. Courant and D. Hilbert

1. *Methods of Mathematical Physics*, Vol. II, Interscience, New York, 1962.

G. Darboux

1. *Lecons sur la théorie générale des surfaces*, Première partie, Gauthier-Villars, Paris (nouveau tirage) 1941.

E. de Giorgi

1. Una extensione del teorema di Bernstein, *Ann. Scuola Norm. Sup. Pisa 19* (1965), 79-85.

E. de Giorgi and G. Stampacchia

   1. Sulle singolarità eliminabili delle ipersuperficie minimali, *Atti Accad. Naz. Lincei, Rend. Cl. Sci. Fis. Mat. Natur.* (ser 8) *38* (1965), 352-357.

J. Douglas

   1. Solution of the problem of Plateau, *Trans. Amer. Math. Soc.,* *33* (1931), 263-321.

   2. Minimal surfaces of higher topological structure, *Ann. of Math.,* (2) *40* (1939), 205-298.

H. Federer and W. H. Fleming

   1. Normal and integral currents, *Ann. of Math.,* 72 (1960), 458-520.

R. Finn

   1. Isolated singularities of solutions of non-linear partial differential equations, *Trans. Amer. Math. Soc.,* *75* (1953), 383-404.

   2. On equations of minimal surface type, *Ann. of Math.* (2) *60* (1954), 397-416.

   3. On a problem of type, with application to elliptic partial differential equations, *J. Rational Mech. Anal.,* *3* (1954), 789-799.

   4. New estimates for equations of minimal surface type, *Arch. Rational Mech. Anal.* *14* (1963), 337-375.

   5. Remarks relevant to minimal surfaces and to surfaces of prescribed mean curvature, *J. d'Anal. Math.,* *14* (1965), 139-160.

   6. On a class of conformal metrics, with application to differential geometry in the large, *Comment. Math. Helv.,* *40* (1965), 1-30.

R. Finn and R. Osserman

 1. On the Gauss curvature of non-parametric minimal surfaces,
    J. Analyse Math., *12* (1964), 351-364.

W. H. Fleming

 1. An example in the problem of least area, *Proc. Amer. Math.
    Soc.*, 7 (1956), 1063-1074.

 2. On the oriented Plateau problem, *Rend. Circ. Mat. Palermo*
    (2) *11* (1962), 69-90.

T. Frankel

 1. On the fundamental group of a compact minimal submanifold,
    Ann. of Math., *83* (1966), 68-73.

P. R. Garabedian

 1. *Partial Differential Equations*, Wiley, New York, 1964.

A. Gray

 1. Minimal varieties and almost Hermitian submanifolds, *Michigan Math. J.*, *12* (1965), 273-287.

E. Heinz

 1. Über die Lösungen der Minimalflächengleichung, *Nachr.
    Akad. Wiss. Göttingen Math.*, *Phys. Kl. II* (1952), 51-56.

 2. Über die Existenz einer Fläche konstanter mittlere Krümmung
    bei vorgegebener Berandung, *Math. Ann.*, *127* (1954), 258-
    287.

S. Hildebrandt

 1. Über das Randverhalten von Minimalflächen, *Math. Annalen*,
    165 (1966), 1-18.

 2. Über Minimalflächen mit freiem Rand, *Math.
    Zeitschrift*, *95* (1967), 1-19.

E. Hopf

1. On an inequality for minimal surfaces $z = z(x, y)$, *J. Rational Mech. Anal.*, *2* (1953), 519-522; 801-802.

H. Hopf and W. Rinow

1. Über den Begriff der vollständigen differentialgeometrischen Fläche, *Comment. Math. Helv.*, *3* (1931), 209-225.

Wu-Yi Hsiang

1. On the compact homogeneous minimal submanifolds, *Proc. Nat'l. Acad. Sci.*, *56* (1966), 5-6.

2. Remarks on closed minimal submanifolds in the standard Riemannian m-sphere, *J. Diff. Geom. 1* (1967), 257-267.

A. Huber

1. On subharmonic functions and differential geometry in the large, *Comment. Math. Helv.*, *32* (1957), 13-72.

H. Jenkins

1. On two-dimensional variational problems in parametric form, *Arch. Rational Mech. Anal.*, *8* (1961), 181-206.

2. On quasi-linear elliptic equations which arise from variational problems, *J. Math. Mech.*, *10* (1961), 705-728.

3. Super solutions for quasi-linear elliptic equations, *Arch. Rational Mech. Anal.*, *16* (1964), 402-410

H. Jenkins and J. Serrin

1. Variational problems of minimal surface type I, *Arch. Rational Mech. Anal.*, *12* (1963), 185-212.

2. Variational problems of minimal surface type II: boundary value problems for the minimal surface equation, *Arch. Rational Mech. Anal.*, *21* (1965/66), 321-342

3. The Dirichlet problem for the minimal surface equation in higher dimensions, *J. Reine Angew, Math.*, *229* (1968), 170-187.

L. Jonker

1. A theorem on minimal surfaces, *J. Differential Geometry, 3* (1969), 351-360.

K. Jörgens

1. Über die Lösungen der Differentialgleichung $rt - s^2 = 1$, *Math. Ann., 127* (1954), 130-134.

T. Klotz and R. Osserman

1. On complete surfaces in E^3 with constant mean curvature, *Comment. Math. Helv., 41* (1966-67), 313-318.

T. Klotz and L. Sario

1. Existence of complete minimal surfaces of arbitrary connectivity and genus, *Proc. Nat. Acad. Sci., 54* (1965), 42-44.

H. B. Lawson, Jr.

1. Local rigidity theorems for minimal hypersurfaces, *Annals of Math.*, 89 (1969), 187-197.

2. Stanford Thesis, 1968.

3. Compact minimal surfaces in S^3, pp. 275-282 of *Global Analysis*, Proc. Symp. Pure Math., Vol. XV, Amer. Math. Soc., Providence, R.I., 1970.

4. Some intrinsic characterizations of minimal surfaces, *J. d'Analyse Math.*, 24 (1971), 151-161.

P. Lévy

1. Le problème de Plateau, *Mathematica, 23* (1948), 1-45.

H. Lewy

1. A priori limitations for solutions of Monge-Ampère equations, II, *Trans. Amer. Soc., 41* (1937), 365-374.

2. Aspects of the Calculus of Variations, (Notes by J. W. Green after lectures by Hans Lewy) Univ. of California Press, Berkeley 1939.

Ü. Lumiste

  1. On the theory of two-dimensional minimal surfaces: I, II, III, IV, *Tartu Rikl. Ül. Toimetised, 102* (1961), 3-15, 16-28; *129* (1962), 74-89, 90-102.

G. R. MacLane

  1. On asymptotic values, Abstract 603-166, *Notices Amer. Math. Soc., 10* (1963), 482-483.

M. Miranda

  1. Un teorema di esistenza e unicità per il problema del'area minima in *n* variabili, *Ann. Scuola Normale Sup. Pisa* (Ser 3) *19* (1965), 233-250.

C. B. Morrey

  1. The problem of Plateau on a Riemannian manifold, *Ann. of Math.* (2) *49* (1948), 807-851.

  2. The higher-dimensional Plateau problem on a Riemannian manifold, *Proc. Nat. Acad. Sci. U.S.A., 54* (1965), 1029-1035.

  3. *Multiple Integrals in the Calculus of Variations*, New York, Springer-Verlag, 1966.

J. Moser

  1. On Harnack's Theorem for elliptic differential equations, *Comm. Pure Appl. Math., 14* (1961), 577-591.

C. H. Müntz

  1. Die Lösung des Plateauschen Problems über konvexen Bereichen, *Math. Ann., 94* (1925), 53-96.

J. C. C. Nitsche

  1. Über eine mit der Minimalflächengleichung zusammenhängende analytische Funktion und den Bernsteinschen Satz., *Arch. Math., 7* (1956), 417-419.

2. Elementary proof of Bernstein's theorem on minimal surfaces, *Ann. of Math.*, (2) *66* (1957), 543-544.

3. A characterization of the catenoid, *J. Math. Mech., 11* (1962), 293-302.

4. On new results in the theory of minimal surfaces, *Bull. Amer. Math. Soc., 71* (1965), 195-270.

5. Über ein verallgemeinertes Dirichletsches Problem für die Minimalflächengleichung und hebbare Unstetigkeiten ihrer Lösungen, *Math. Ann., 158* (1965), 203-214.

6. On the non-solvability of Dirichlet's problem for the minimal surface equation, *J. Math. Mech., 14* (1965), 779-788.

7. Contours bounding at least three solutions of Plateau's problem, *Arch. Rat'l. Mech. Anal., 30* (1968), 1-11.

8. Note on the nonexistence of minimal surfaces, *Proc. Amer. Math. Soc., 19* (1968), 1303-1305.

Johannes Nitsche and Joachim Nitsche

1. Über reguläre Variationsprobleme, *Rend. Circ. Mat. Palermo* (2) *8* (1959), 346-353.

R. Osserman

1. Proof of a conjecture of Nirenberg, *Comm. Pure Appl. Math., 12* (1959), 229-232.

2. On the Gauss curvature of minimal surfaces, *Trans. Amer. Math. Soc., 96* (1960), 115-128.

3. Minimal surfaces in the large, *Comment. Math. Helv., 35* (1961), 65-76.

4. On complete minimal surfaces, *Arch. Rational Mech. Anal., 13* (1963), 392-404.

5. Global properties of minimal surfaces in E^3 and E^n, *Ann. of Math.*, (2) *80* (1964), 340-364.

6. Global properties of classical minimal surfaces, *Duke Math. J., 32* (1965), 565-573.

7. Le théorème de Bernstein pour des systèmes, *C. R. Acad. Sci. Paris, 262* (1966), sér A, 571-574.

8. Minimal surfaces (in Russian), *Uspekhi Mat. NAUK, 22* (1967), 56-136.

9. Minimal varieties, *Bull. Amer. Math. Soc.*, 75 (1969), 1092-1120.

10. Some properties of solutions to the minimal surface system for arbitrary codimension, pp. 283-291 of *Global Analysis*, Proc. Symp. Pure Math., Vol. XV, Amer. Math. Soc., Providence, R.I., 1970.

T. Otsuki

1. Minimal hypersurfaces in a Riemannian manifold of constant curvature, *Amer. J. Math.*, 92 (1970), 145-173.

M. Pinl

1. B-Kugelbilder reeler Minimalflächen in R^4, *Math. Z., 59* (1953), 290-295.

2. Minimalflächen fester Gausscher Krümmung, *Math. Ann., 136* (1958), 34-40.

3. Über einen Satz von G. Ricci-Curbastro und die Gaussche Krümmung der Minimalflächen, II, *Arch. Math., 15* (1964), 232-240.

T. Radó

1. Über den analytischen Character der Minimalflächen, *Math. Z., 24* (1925), 321-327.

2. The problem of the least area and the problem of Plateau, *Math. Z., 32* (1930), 763-796.

3. On the problem of Plateau, *Ergebnisse der Mathematik und ihrer Grenzgebiete*, Springer-Verlag, Berlin, 1933.

E. R. Reifenberg

1. Solution of the Plateau problem for m-dimensional surfaces of varying topological type, *Acta. Math.*, *104* (1960), 1-92.
2. An epiperimetric inequality related to the analyticity of minimal surfaces, *Ann. of Math.*, *80* (1964), 1-14.
3. On the analyticity of minimal surfaces, *Ann. of Math.*, *80* (1964), 15-21.

R. C. Reilly

1. Extrinsic rigidity theorems for compact submanifolds of the sphere, *J. Diff. Geom.*, 4 (1970), 487-497.

L. Sario and K. Noshiro

1. *Value Distribution Theory, Appendix II: Mapping of arbitrary minimal surfaces*, Van Nostrand, 1966.

A. H. Schoen

1. Infinite regular warped polyhedra and infinite periodic minimal surfaces, Abstract 658–30, *Notices Amer. Math. Soc.*, *15* (1968), p. 727.

J. Serrin

1. A priori estimates for solutions of the minimal surface equation, *Arch. Rational Mech. Anal.*, *14* (1963), 376-383.
2. Removable singularities of elliptic equations, II, *Arch Rational Mech. Anal.*, *20* (1965), 163-169.
3. The Dirichlet problem for quasilinear equations with many independent variables, *Proc. Nat'l. Acad. Sci.*, *58* (1967), 1824-1835.

J. Simons

1. Minimal varieties in Riemannian manifolds, *Ann. of Math.*, *88* (1968), 62-105.

D. J. Struik

1. *Mehrdimensionalen Differentialgeometrie*, Berlin, Springer-Verlag, 1922.

T. Takahashi

1. Minimal immersions of Riemannian manifolds, *J. Math. Soc., Japan, 18* (1966), 380-385.

M. Takeuchi and S. Kobayashi

1. Minimal embeddings of R-spaces, *J. Differential Geometry, 2* (1968), 203-215.

K. Voss

1. Über vollständige Minimalflächen, *L'Enseignement Math., 10* (1964), 316-317.

H. Werner

1. Das Problem von Douglas für Flächen konstanter mittlerer Krümmung, *Math. Ann., 133* (1957), 303-319.

H. Weyl

1. Meromorphic functions and analytic curves, *Annals of Math. Studies, No. 12*, Princeton University Press, Princeton, N. J., 1943.

ADDITIONAL REFERENCES

Books and Survey Articles

MSG

Minimal Submanifolds and Geodesics, Proceedings of the Japan–United States Seminar on Minimal Submanifolds, including Geodesics, Tokyo, 1977, Kagai Publications, Tokyo, 1978.

SMS

Seminar on Minimal Submanifolds, edited by Enrico Bombieri, *Ann. of Math. Studies*, 103, Princeton, 1983.

W. K. Allard and F. J. Almgren, Jr. eds.

I. *Geometric Measure Theory and the Calculus of Variations*, Proceedings of Symposia in Pure Mathematics, *Vol. 44*, Amer. Math. Soc., Providence, 1986.

F. J. Almgren, Jr.

I. Minimal surface forms, *Math. Intelligencer, 4* (1982), 164-172.

F. J. Almgren, Jr., and J. E. Taylor

I. The geometry of soap films and soap bubbles, *Scientific American*, July 1976, pp. 82-93.

J. L. M. Barbosa and A. G. Colares

I. Examples of Minimal Surfaces in $\mathbb{R}^3$ (to appear).

R. Böhme

I. New results on the classical problem of Plateau on the existence of many solutions, *Séminaire Bourbaki*, 34e année, *no. 579* (1981/2), 1-20.

E. Bombieri

I. Recent progress in the theory of minimal surfaces, *L'Enseignement Math., 25* (1979), 1-8.

II. An Introduction to Minimal Currents and Parametric Variational Problems, *Mathematical Reports, Vol. 2, Part 3*, Harwood, London, 1985.

E. de Giorgi, F. Colombini, and L. C. Piccinini
I. *Frontiere orientate di misura minima e questioni collegate*, Scuola Normale Superiore, Pisa, 1972.

M. P. do Carmo
I. Stability of minimal submanifolds, pp. 129-139 of *Global Differnetial Gometry and Global Analysis*, Lecture Notes in Mathematics, *Vol. 838*, Springer-Verlag, Berlin, 1981.

H. Federer
I. *Geometric Measure Theory*, Springer-Verlag, Berlin, 1969.

A. T. Fomenko
I. *Topological Variational Problems* (in Russian), Moscow University, 1984.

E. Giusti
I. *Minimal Surfaces and Functions of Bounded Variation*, Birkhäuser, Boston, 1984.

R. Harvey and H. B. Lawson, Jr.
I. Geometries associated to the group SU_n and varieties of minimal submanifolds arising from the Cayley arithmetic, pp. 43-59 of *MSG*.

S. Hildebrandt and A. Tromba
I. *Mathematics and Optimal Form*, Scientific American Books, New York, 1985.

H. B. Lawson, Jr.
I. Lectures on Minimal Surfaces, *IMPA*, Rio de Janeiro, 1970; 2d ed.: Publish or Perish Press, Berkeley, 1980.

II. *Minimal Varieties in Real and Complex Geometry*, Univ. of Montreal Press, Montreal, 1973.

III. Surfaces minimales et la construction de Calabi-Penrose, *Séminaire Bourbaki*, 36e année, *no. 624* (1983/84), 1-15.

U. Massari and M. Miranda

   I. *Minimal Surfaces of Codimension One*, North Holland Mathematical Studies, *91*, North-Holland, Amsterdam, 1984.

W. H. Meeks III

   I. *Lectures on Plateau's Problem*, Escola de Geometria Diferencial, Universidade Federal do Ceará (Brazil), De 17 a 28 de Julho de 1978.

  II. A survey of the geometric results in the classical theory of minimal surfaces, *Bol. Soc. Bras. Mat.*, *12* (1981), 29-86.

J. C. C. Nitsche

   I. Plateau's Problems and Their Modern Ramifications, *Amer. Math. Monthly*, *81* (1974), 945-968.

  II. *Vorlesungen über Minimalflächen*, Springer-Verlag, Berlin, 1975.

 III. Uniqueness and non-uniqueness for Plateau's problem—one of the last major questions, pp. 143-161 of *MSG*.

 IV. *Minimal Surfaces and Partial Differential Equations*, MAA Studies in Mathematics, *Vol. 23*, Mathematical Association of America, Washington, D.C., 1982, pp. 69-142.

R. Osserman

   I. Branched immersions of surfaces, *INDAM Symposia Math.*, *10* (1972), 141-158.

  II. Properties of solutions to the minimal surface equation in higher codimension, pp. 163-172 of *MSG*.

III. The isoperimetric inequality, *Bull. Amer. Math. Soc., 84* (1978), 1182-1238.

IV. Minimal surfaces, Gauss maps, total curvature, eigenvalue estimates, and stability, pp. 199-227 of *The Chern Symposium 1979*, Springer-Verlag, New York, 1980.

V. The minimal surface equation, pp. 237-259 of *Seminar on Nonlinear Partial Differential Equations*, Mathematical Sciences Research Institute Publications, *Vol. 2*, Springer-Verlag, New York, 1984.

I. Peterson

I. Three bites in a doughnut; computer-generated pictures contribute to the discovery of a new minimal surface, *Science News, 127, No. 11* (1985), 161-176.

J. T. Pitts

I. *Existence and Regularity of Minimal Surfaces on Riemannian manifolds*, Princeton Univ. Press, Princeton, 1981.

L. Simon

I. Survey Lectures on Minimal Submanifolds, pp. 3-52 of *SMS*.

II. *Lectures on Geometric Measure Theory*, Centre for Mathematical Analysis, Australian National University, 1984.

G. Toth

I. *Harmonic and Minimal Maps: With Applications in Geometry and Physics*, Ellis Horwood, Chichester, England, 1984.

Research Papers

F. J. Almgren, Jr.

4. Optimal isoperimetric inequalities, *Bull. Amer. Math. Soc., 13* (1985), 123-126.

F. J. Almgren, Jr., and L. Simon

1. Existence of embedded solutions of Plateau's problem, *Ann. Scuola Norm. Sup. Pisa, 6* (1979), 447-495.

F. J. Almgren, Jr., and W. P. Thurston

1. Examples of unknotted curves which bound only surfaces of high genus within their convex hulls, *Ann. of Math., (2) 105* (1977), 527-538.

H. W. Alt

1. Verzweigungspunkte von H-Flächen, I.: *Math. Z., 127* (1972), 333-362; II: *Math. Ann., 201* (1973), 33-55.

Ju. A. Aminov

1. On the instability of a minimal surface in an n-dimensional Riemannian space of constant curvature, *Math. USSR-Sbornik, 29* (1976), 359-375.

M. T. Anderson

1. Complete minimal varieties in hyperbolic space, *Invent. Math., 69* (1982), 447-494.

2. Curvature estimates for minimal surfaces in 3-manifolds, *Ann. Sci. École Norm. Sup., 18* (1985), 89-105.

3. The compactification of a minimal submanifold in euclidean space by its Gauss map (to appear).

J. L. M. Barbosa

1. An extrinsic rigidity theorem for minimal immersions from S^2 into S^n, *J. Differential Geometry, 14* (1979), 355-368.

J. L. M. Barbosa and M. do Carmo

1. On the size of a stable minimal surface in R^3, *Amer. J. Math., 98* (1976), 515-528.

2. Stability of minimal surfaces and eigenvalues of the Laplacian, *Math. Z., 173* (1980), 13-28.

3. Stability of minimal surfaces in spaces of constant curvature, *Bol. Soc. Brasil. Mat., 11* (1980), 1-10.

4. A necessary condition for a metric in M^n to be minimally immersed in $\mathbb{R}^{n+1}$, *An. Acad. Brasil. Ci., 50* (1978), 451-454.

5. Stability of hypersurfaces with constant mean curvature, *Math. Z., 185* (1984), 339-353.

M. Beeson

1. On interior branch points of minimal surfaces, *Math. Z., 171* (1980), 133-154.

2. Some results on finiteness in Plateau's problem, Part I: *Math Z., 175* (1980), 103-123; Part II: *Math. Z., 181* (1982), 1-30.

R. Böhme and F. Tomi

1. Zur Struktur der Lösungsmenge des Plateauproblems, *Math. Z., 133* (1973), 1-29.

R. Böhme and A. J. Tromba

1. The index theorem for classical minimal surfaces, *Ann. of Math., 113* (1981), 447-499.

E. Bombieri and L. Simon

1. On the Gehring link problem, in *SMS*, pp. 271-274.

R. Bryant

1. Conformal and minimal immersions of compact surfaces into the 4-sphere, *J. Differential Geom., 17* (1982), 455-473.

2. A duality theorem for Willmore surfaces, *J. Differential Geom., 20* (1984), 23-53.

G. D. Chakerian

1. The isoperimetric theorem for curves on minimal surfaces, *Proc. Amer. Math. Soc., 69* (1978), 312-313.

I. Chavel

    1. On A. Hurwitz' method in isoperimetric inequalities, *Proc. Amer. Math. Soc., 71* (1978), 275-279.

C. C. Chen

    1. Complete minimal surfaces with total curvature -2π, *Bol. Soc. Brasil. Mat., 10* (1979), 71-76.

    2. A characterization of the catenoid, *An. Acad. Brasil. Ci., 51* (1979), 1-3.

    3. Elliptic functions and non-existence of complete minimal surfaces of certain type, *Proc. Amer. Math. Soc., 79* (1980), 289-293.

    4. On the image of the generalized Gauss Map of a complete minimal surface in $\mathbb{R}^4$, *Pacfic J. Math., 102* (1982), 9-14.

C. C. Chen and F. Gackstatter

    1. Elliptische und hyperelliptische Funktionen und vollständige Minimalflächen vom Enneperschen Typ, *Math. Ann., 259* (1982), 359-369.

S.-Y. Cheng, P. Li, and S.-T. Yau

    1. Heat equations on minimal submanifolds and their applications, *Amer. J. Math., 106* (1984), 1033-1065.

S.-S. Chern and S. I. Goldberg

    1. On the volume decreasing property of a class of real harmonic mappings, *Amer. J. Math., 97* (1975), 133-147.

S. S. Chern and R. Osserman

    2. Remarks on the Riemannian metric of a minimal submanifold, pp. 49-90 of *Geometry Symposium, Utrecht 1980*, Lecture Notes in Math., *Vol. 894*, Springer-Verlag, Berlin, 1981.

S. S. Chern and J. G. Wolfson

    1. Minimal surfaces by moving frames, *Amer. J. Math., 105* (1983), 59-83.

H. I. Choi, W. H. Meeks, and B. White

  1. A rigidity theorem for properly embedded minimal surfaces in $\mathbb{R}^3$ (to appear).

H. I. Choi and R. Schoen

  1. The space of minimal embeddings of a surface into a three-dimensional manifold of positive Ricci curvature (to appear).

C. J. Costa

  1. Example of a complete minimal immersion in $\mathbb{R}^3$ of genus one and three embedded ends (to appear). *where ?*

A. M. da Silveira

  1. Stability of complete noncompact surfaces with constant mean curvature (to appear). *where*

A. M. Din and W. J. Zakrzewski

  1. General classical solutions in the $\mathbb{C}P^{n-1}$ model, *Nuclear Phys., B. 174* (1980), 397-406.

  2. Properties of the general classical $\mathbb{C}P^{n-1}$ model, *Phys. Lett., 95B* (1980), 419-422.

M. do Carmo and C. K. Peng

  1. Stable complete minimal surfaces in $\mathbb{R}^3$ are planes, *Bull. Amer. Math. Soc., 1* (1979), 903-906.

J. Eells

  1. Minimal graphs, *Manuscripta Math., 28* (1979), 101-108.

J. Eells and L. Lemaire

  1. A report on harmonic maps, *Bull. London Math. Soc., 10* (1978), 1-68.

  2. On the construction of harmonic and holomorphic maps between surfaces, *Math. Ann., 252* (1980), 27-52.

J. Eells and J. H. Sampson

  1. Harmonic mappings of Riemannian manifolds, *Amer. J. Math., 86* (1964), 109-160.

pg154

pg147

J. Eells and J. C. Wood

   1. Harmonic maps from surfaces to complex projective spaces, *Advances in Math.*, *49* (1983), 217-263.

H. Federer

   1. Some theorems on integral currents, *Trans. Amer. Math. Soc.*, *117* (1965), 43-67.

J. Feinberg

   1. The isoperimetric inequality for doubly-connected minimal surfaces in $\mathbb{R}^n$, *J. Analyse Math.*, *32* (1977), 249-278.

D. Fischer-Colbrie

   1. Some rigidity theorems for minimal submanifolds of the sphere, *Acta Math.*, *145* (1980), 29-46.

   2. On complete minimal surfaces with finite Morse index in three manifolds, *Inventiones Math.* (to appear).

D. Fischer-Colbrie and R. Schoen

   1. The structure of complete stable minimal surfaces in 3-manifolds of non-negative scalar curvature, *Comm. Pure Appl. Math.*, *33* (1980), 199-211.

T. Frankel

   2. Applications of Duschek's formula to cosmology and minimal surfaces, *Bull. Amer. Math. Soc.*, *81* (1975), 579-582.

T. Frankel and G. J. Galloway

   1. Stable minimal surfaces and spatial topology in general relativity, *Math. Z.*, *181* (1982), 395-406.

M. Freedman, J. Hass, and P. Scott

   1. Least area incompressible surfaces in 3-manifolds, *Invent. Math.*, *71* (1983), 609-642.

H. Fujimoto

   1. On the Gauss map of a complete minimal surface in $\mathbb{R}^m$, *J. Math. Soc. Japan*, *35* (1983), 279-288.

2. Value distribution of the Gauss maps of complete minimal surfaces in $\mathbb{R}^m$, *J. Math. Soc. Japan*, *35* (1983), 663-681.

F. Gackstatter

1. Über abelsche Minimalflächen, *Math. Nachr.*, *74* (1976), 157-165.

2. Über die Dimension einer Minimalfläche und zur Ungleichung von St. Cohn-Vossen, *Arch. Rational Mech. Anal.*, *61* (1976), 141-152.

F. Gackstatter and R. Kunert

1. Konstruktion vollständiger Minimalflächen von endlicher Gesamtkrümmung, *Arch. Rational Mech. Anal*, *65* (1977), 289-297.

M. Gage

1. A proof of Gehring's linked spheres conjecture, *Duke Math. J.*, *47* (1980), 615-620.

V. Glaser and R. Stora

1. Regular solutions of the $\mathbb{C}P^n$ models and further generalizations (*CERN* preprint), 1980.

C. C. Goes and P. A. Q. Simoes

1. Some remarks on minimal immersions in the hyperbolic spaces (to appear).

M. Gromov

1. Filling Riemannian manifolds, *J. Differential Geom.*, *18* (1983), 1-147.

M. Gromov and H. B. Lawson, Jr.

1. Positive scalar curvature and the Dirac operator on complete Riemannian manifolds, *I.H.E.S. Publ. Math.*, *No. 58* (1983), 83-196.

R. D. Gulliver

1. Regularity of minizing surfaces of prescribed mean curvature, *Ann. of Math.*, *97* (1973), 275-305.

2. Branched immersions of surfaces and reduction of topological type, I: *Math. Z.*, *145* (1975), 267-288; II: *Math. Ann.*, *230* (1977), 25-48.

3. Index and total curvature of complete minimal surfaces (to appear).

R. D. Gulliver and H. B. Lawson, Jr.

1. The structure of stable minimal hypersurfaces near a singularity (to appear).

R. D. Gulliver, R. Osserman, and H. L. Royden

1. A theory of branched immersions of surfaces, *Amer. J. Math.*, *95* (1973), 750-812.

R. D. Gulliver and J. Spruck

1. On embedded minimal surfaces, *Ann. of Math.*, *103* (1976), 331-347, with a correction in *Ann. of Math.*, *109* (1979), 407-412.

A. Haeflinger

1. Some remarks on foliations with minimal leaves, *J. Differential Geom.*, *15* (1980), 269-284.

P. Hall

1. Two topological examples in minimal surface theory, *J. Differential Geom.*, *19* (1984), 475-481.

R. Hardt and L. Simon

1. Boundary regularity and embedded solutions for the oriented Plateau problem, *Ann. of Math.*, *110* (1979), 439-486.

R. Harvey and H. B. Lawson, Jr.

1. Extending minimal varieties, *Invent. Math.*, *28* (1975), 209-226.

2. A constellation of minimal varieties defined over the group G_2, pp. 167-187 of *Partial Differential Equations and Geometry*, Proceedings of Park City Conference, edited by C. I. Byrnes, M. Dekker, New York, 1979.

3. Calibrated geometries, *Acta Math.*, *148* (1982), 47-157.

4. Calibrated foliations, *Amer. J. Math.*, *103* (1981), 411-435.

J. Hass

1. The geometry of the slice-ribbon problem, *Math. Proc. Cambridge Phil. Soc.*, *94* (1983), 101-108.

2. Complete area minimizing minimal surfaces which are not totally geodesic, *Pacific J. Math.*, *11* (1984), 35-38.

3. Group actions on 3-manifolds with non-Haken quotients and intersections of surfaces minimizing area in their homology class (to appear).

S. Hildebrandt

3. Boundary behavior of minimal surfaces, *Arch. Rational Mech. Anal.*, *35* (1969), 47-82.

S. Hildebrandt, J. Jost, and K. O. Widman

1. Harmonic mappings and minimal submanifolds, *Invent. Math.*, *62* (1980), 269-298.

S. Hildebrandt and J. C. C. Nitsche

1. A uniqueness theorem for surfaces of least area with partially free boundaries on obstacles, *Arch. Rational Mech. Anal.*, *79* (1982), 189-218.

D. Hoffman and W. H. Meeks III

1. A complete embedded minimal surface in $\mathbb{R}^3$ with genus one and three ends, *J. Differential Geom.*, *21* (1985), 109-127.

2. (In preparation.)

D. Hoffman and R. Osserman

1. The geometry of the generalized Gauss map, *Mem. Amer. Math. Soc., No. 236*, 1980.

2. The area of the generalized Gaussian image and the stability of minimal surfaces in S^n and $\mathbb{R}^n$, *Math. Ann., 260* (1982), 437-452.

3. The Gauss map of surfaces in $\mathbb{R}^n$, *J. Differential Geom., 18* (1983), 733-754.

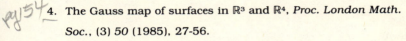 4. The Gauss map of surfaces in $\mathbb{R}^3$ and $\mathbb{R}^4$, *Proc. London Math. Soc., (3) 50* (1985), 27-56.

D. Hoffman, R. Osserman, and R. Schoen

1. On the Gauss map of complete surfaces of constant mean curvature in $\mathbb{R}^3$ and $\mathbb{R}^4$, *Comment. Math. Helv., 57* (1982), 519-531.

Wu-Yi Hsiang

3. Generalized rotational hypersurfaces of constant mean curvature in the euclidean spaces I., *J. Differential Geom., 17* (1982), 337-356.

4. New examples of minimal imbedding of S^{n-1} into $S^n(1)$—the spherical Bernstein problem for $n=4,5,6$, *Bull. Amer. Math. Soc., 7* (1982), 377-379.

Wu-Yi Hsiang and H. B. Lawson, Jr.

1. Minimal submanifolds of low cohomogeneity, *J. Differential Geom., 5* (1971), 1-38. (Corrections in F. Uchida, *J. Differential Geom., 15* (1980), 569-574.)

L. P. M. Jorge and W. H. Meeks III

1. The topology of complete minimal surfaces of finite total curvature, *Topology, 22* (1983), 203-221.

L. P. M. Jorge and F. Xavier

1. A complete minimal surface in $\mathbb{R}^3$ between two parallel planes, *Ann. of Math.*, *112* (1980), 203-206.

F. W. Kamber and P. Tondeur

1. Harmonic foliations, in *Harmonic Maps*, Lecture Notes in Math. *949*, Springer-Verlag, Berlin, 1982.

A. Kasue

1. A gap theorem for minimal submanifolds of Euclidean space (to appear).

S. Kawai

1. A theorem of Bernstein type for minimal surfaces in $\mathbb{R}^4$, *Tôhoku Math. J.*, *36* (1984), 377-384.

K. Kenmotsu

1. Weierstrass formula for surfaces of prescribed mean curvature, *Math. Ann.*, *245* (1979), 89-99.

M. Koiso

1. On the finite solvability of Plateau's problem for extreme curves, *Osaka J. Math.*, *20* (1983), 177-183.

H. B. Lawson, Jr.

5. The global behavior of minimal surfaces in S^n, *Ann. of Math.*, *92* (1970), 224-237.

6. Complete minimal surfaces in S^3, *Ann. of Math.*, *92* (1970), 335-374.

7. The equivariant Plateau problem and interior regularity, *Trans. Amer. Math. Soc.*, *173* (1972), 231-249.

H. B. Lawson, Jr., and R. Osserman

1. Non-existence, non-uniqueness and irregularity of solutions to the minimal surface system, *Acta Math.*, *139* (1977), 1-17.

H. B. Lawson, Jr., and J. Simons

1. On stable currents and their applications to global problems in real and complex geometry, *Ann. of Math.*, *98* (1973), 427-450.

P. Li, R. Schoen, and S.-T. Yau

1. On the isoperimetric inequality for minimal surfaces (to appear).

W. H. Meeks III

1. The conformal structure and geometry of triply periodic minimal surfaces in $\mathbb{R}^3$, Ph.D. thesis, Berkeley, 1975.

2. The conformal structure and geometry of triply periodic minimal surfaces in $\mathbb{R}^3$, *Bull. Amer. Math. Soc.*, *83* (1977), 134-136.

3. The classification of complete minimal surfaces in $\mathbb{R}^3$ with total curvature greater than -8π, *Duke Math. J.*, *48* (1981), 523-535.

4. Uniqueness theorems for minimal surfaces, *Illinois J. Math.*, *25* (1981), 318-336.

W. H. Meeks III, L. Simon, and S.-T. Yau

1. Embedded minimal surfaces, exotic spheres, and manifolds with positive Ricci curvature, *Ann. of Math.*, *116* (1982), 621-659.

W. H. Meeks III and S.-T. Yau

1. The classical Plateau problem and the topology of three dimensional manifolds, *Topology.* *21* (1982), 409-440.

2. Topology of three dimensional manifolds and the embedding problems in minimal surface theory, *Ann. of Math.*, *112* (1980), 441-485.

3. The equivariant Dehn's lemma and loop theorem, *Comment. Math. Helv.*, *56* (1981), 225-239

4. The existence of embedded minimal surfaces and the problem of uniqueness, *Math. Z.*, *179* (1982), 151-168.

5. The equivariant loop theorem for three-dimensional manifolds

and a review of existence theorems for minimal surfaces, pp. 153-163 of *The Smith Conjecture*, Academic Press, New York, 1984.

M. J. Micallef

1. Stable minimal surfaces in Euclidean space, *J. Differential Geom.*, *19* (1984), 57-84.

2. Stable minimal surfaces in flat tori (to appear).

T. K. Milnor

1. Harmonically immersed surfaces, *J. Differential Geom.*, *14* (1979), 205-214.

C. W. Misner

1. Harmonic maps as models for physical theories, *Phys. Rev.*, *D 18* (1978), 4510-4525.

C. W. Misner, K. S. Thorne, and J. A. Wheeler

1. *Gravitation*, W. H. Freeman, San Francisco, 1973.

J. D. Moore

1. On stability of minimal spheres and a two-dimensional version of Synge's Theorem, *Arch. Math.*, *44* (1985), 278-281.

F. Morgan

1. A smooth curve in $\mathbb{R}^4$ bounding a continuum of area minimizing surfaces, *Duke Math. J.*, *43* (1976), 867-870.

2. Almost every curve in $\mathbb{R}^3$ bounds a unique area minimizing surface, *Invent. Math.*, *45* (1978), 253-297.

3. A smooth curve in $\mathbb{R}^3$ bounding a continuum of minimal manifolds, *Arch. Rational Mech. Anal.*, *75* (1981), 193-197.

4. On the singular structure of two-dimensional area minimizing surfaces in $\mathbb{R}^n$, *Math. Ann.*, *261* (1982), 101-110.

5. On finiteness of the number of stable minimal hypersurfaces with a fixed boundary, *Bull. Amer. Math. Soc.*, *13* (1985), 133-136.

H. Mori

    1. Remarks on the size of a stable minimal surface in a Riemannian manifold (to appear).

T. Nagano and B. Smyth

    1. Periodic minimal surfaces and Weyl groups, *Acta Math.*, *145* (1980), 1-27.

N. Nakauchi

    1. Multiply connected minimal surfaces and the geometric annulus theorem, *J. Math. Soc. Japan*, *37* (1985), 17-39.

J. C. C. Nitsche

    9. A new uniqueness theorem for minimal surfaces, *Arch. Rational Mech. Anal.*, *52* (1973), 319-329.

    10. Non-uniqueness for Plateau's problem. A bifurcation process, *Ann. Acad. Sci. Fenn. Ser. A I Math*, *2* (1976), 361-373.

M. E. G. G. de Oliveira

    1. Non-orientable minimal surfaces in $\mathbb{R}^n$ (to appear).

R. Osserman

    11. A proof of the regularity everywhere of the classical solution to Plateau's problem, *Ann. of Math.*, *91* (1970), 550-569.

    12. On Bers' theorem on isolated singularities, *Indiana Univ. Math. J.*, *23* (1973), 337-342.

    13. Some remarks on the isoperimetric inequality and a problem of Gehring, *J. Analyse Math.*, *30* (1976), 404-410.

    14. The total curvature of algebraic surfaces, pp. 249-257 of *Contributions to Analysis and Geometry*, John Hopkins University Press, Baltimore, 1982.

R. Osserman and M. Schiffer

    1. Doubly-connected minimal surfaces, *Arch. Rational Mech. Anal.*, *58* (1975), 285-307.

R. S. Palais and C.-L. Terng

    1. Reduction of variables for minimal submanifolds (to appear).

B. Palmer

    1. Ph.D. thesis, Stanford, 1985.

H. R. Parks

    1. Explicit determination of area minimizing hypersurfaces, *Duke Math. J.*, *44* (1977), 519-534.

C.-K. Peng

    1. Some new examples of minimal surfaces in $\mathbb{R}^3$ and its applications (to appear).

M. Pinl and W. Ziller

    1. Minimal hypersurfaces in spaces of constant curvature, *J. Differential Geom.*, *11* (1976), 335-343.

G. Ricci-Curbastro

    1. *Opere*, Vol. 1, Edizione Cremonese, Rome, 1956, p. 411.

H. Ruchert

    1. A uniqueness result for Enneper's minimal surface, *Indiana Univ. Math. J.*, *30* (1981), 427-431.

E. A. Ruh

    1. Asymptotic behavior of non-parametric minimal hypersurfaces, *J. Differential Geom.*, *4* (1970), 509-513.

E. A. Ruh and J. Vilms

    1. The tension field of the Gauss map, *Trans. Amer. Math. Soc.*, *149* (1970), 569-573.

H. Rummler

    1. Quelques notions simples en géométrie riemannienne et leurs applications aux feuilletages compacts, *Comment. Math. Helv.*, *54* (1979), 224-239.

H. Sacks and K. Uhlenbeck

1. The existence of minimal immersions of the two-sphere, *Ann. of Math.*, *113* (1981), 1-24.

2. Minimal immersions of closed Riemann surfaces, *Trans. Amer. Math. Soc.*, *271* (1982), 639-652.

R. Schoen

1. Estimates for stable minimal surfaces in three dimensional manifolds, pp. 111-126 of *SMS*.

2. Uniqueness, symmetry, and embeddedness of minimal surfaces, *J. Differential Geom.*, *18* (1983), 791-809.

R. Schoen and L. Simon

1. Regularity of stable minimal hypersurfaces, *Comm. Pure Appl. Math.*, *54* (1981), 741-797.

2. Regularity of simply connected surfaces with quasiconformal Gauss map, pp. 127-145 of *SMS*.

R. Schoen, L. Simon, and S.-T. Yau

1. Curvature estimates for minimal hypersurfaces, *Acta Math.*, *134* (1975), 275-288.

R. Schoen and S.-T. Yau

1. Existence of incompressible minimal surfaces and the topology of three dimensional manifolds with non-negative scalar curvature, *Ann. of Math.*, *110* (1979), 127-142

2. On the structure of manifolds with positive scalar curvature, *Manuscripta Math.*, *28* (1979), 159-183.

3. On the proof of the positive mass conjecture in general relativity, *Comm. Math. Phys.*, *65* (1979), 45-76.

4. Proof of the positive mass theorem, II: *Comm. Math. Phys.*, *79* (1981), 231-260.

5. The existence of a black hole due to condensation of matter, *Comm. Math. Phys., 90* (1983), 575-579.

L. Simon

1. Remarks on curvature estimates for minimal hypersurfaces, *Duke Math. J., 43* (1976), 545-553.

2. On a theorem of de Giorgi and Stampacchia, *Math. Z., 155* (1977), 199-204.

3. On some extensions of Bernstein's theorem, *Math. Z., 154* (1977), 265-273.

4. A Hölder estimate for quasiconformal maps between surfaces in Euclidean space, *Acta Math., 139* (1977), 19-51.

5. Equations of mean curvature type in 2 independent variables, *Pacific J. Math., 69* (1977), 245-268.

B. Smyth

1. Stationary minimal surfaces with boundary on a simplex, *Invent. Math., 76* (1984), 411-420.

B. Solomon

1. On the Gauss map of an area-minimizing hypersurface, *J. Differential Geom., 19* (1984), 221-232.

J. Spruck

1. Remarks on the stability of minimal submanifolds of $\mathbb{R}^n$, *Math. Z., 144* (1975), 169-174.

K. Steffen

1. Flächen konstanter mittlerer Krümmung mit vorgegebenen Volumen oder Flächeninhalt, *Arch. Rational Mech. Anal, 49* (1972), 99-128.

2. Isometric inequalities and the problem of Plateau, *Math. Ann., 222* (1976), 97-144.

K. Steffen and H. Wente

1. The non-existence of branch points in solutions to certain classes of Plateau type variational problems, *Math. Z.*, *163* (1978), 211-238.

D. Sullivan

1. A homological characterization of foliations consisting of minimal surfaces, *Comment. Math. Helv.*, *54* (1979), 218-223.

E. Tausch

1. The *n*-dimensional least area problem for boundaries on a convex cone, *Arch. Rational Mech. Anal.*, *75* (1981), 407-416.

F. Tomi

1. On the finite solvability of Plateau's Problem, *Geom. Topol. III Lat. Amer. Sch. Math. Proc. Rio de Janeiro, 1976*, Springer Lecture Notes Math., *597* (1977), 679-695.

F. Tomi and A. J. Tromba

1. Extreme curves bound an embedded minimal surface of the disk type, *Math. Z.*, *158* (1978), 137-145.

A. J. Tromba

1. On the number of simply connected minimal surfaces spanning a curve, *Mem. Amer. Math. Soc.*, *194* (1977).

2. On the Morse number of embedded and non-embedded minimal immersions spanning wires on the boundary of special bodies in $\mathbb{R}^3$, *Math. Z.*, *188* (1985), 149-170.

K. Uhlenbeck

1. Closed minimal surfaces in hyperbolic 3-manifolds, pp. 147-168 of *SMS*.

H. C. Wente

1. A general existence theorem for surfaces of constant mean curvature, *Math. Z.*, *120* (1971), 277-288.

2. Large solutions to the volume constrained problem, *Arch. Rational Mech. Anal., 75* (1980), 59-77.

3. Counterexample to a conjecture of H. Hopf, *Pacific J. Math.* (to appear).

B. White

1. Existence of least area mappings of *N*-dimensional domains, *Ann. of Math., 118* (1983), 179-185.

2. Generic regularity of unoriented two-dimensional area-minimizing surfaces, *Ann. of Math., 121* (1985), 595-603.

3. (In preparation.)

F. Xavier

1. The Gauss map of a complete non-flat minimal surface cannot omit 7 points on the sphere, *Ann. of Math., 113* (1981), 211-214; Erratum: *Ann. of Math., 115* (1982), 667.

2. Convex hulls of complete minimal surfaces, *Math. Ann., 269* (1984), 179-182.

AUTHOR INDEX

201

SUBJECT INDEX

A CATALOG OF SELECTED

DOVER BOOKS

IN ALL FIELDS OF INTEREST

A CATALOG OF SELECTED DOVER BOOKS IN ALL FIELDS OF INTEREST

LASERS AND HOLOGRAPHY, Winston E. Kock. Sound introduction to burgeoning field, expanded (1981) for second edition. 84 illustrations. 160pp. 5⅜ × 8¼. (EUK) 24041-X Pa. $3.50

FLORAL STAINED GLASS PATTERN BOOK, Ed Sibbett, Jr. 96 exquisite floral patterns—irises, poppie, lilies, tulips, geometrics, abstracts, etc.—adaptable to innumerable stained glass projects. 64pp. 8¼ × 11. 24259-5 Pa. $3.50

THE HISTORY OF THE LEWIS AND CLARK EXPEDITION, Meriwether Lewis and William Clark. Edited by Eliott Coues. Great classic edition of Lewis and Clark's day-by-day journals. Complete 1893 edition, edited by Eliott Coues from Biddle's authorized 1814 history. 1508pp. 5⅜ × 8½.
21268-8, 21269-6, 21270-X Pa. Three-vol. set $22.50

ORLEY FARM, Anthony Trollope. Three-dimensional tale of great criminal case. Original Millais illustrations illuminate marvelous panorama of Victorian society. Plot was author's favorite. 736pp. 5⅜ × 8½. 24181-5 Pa. $10.95

THE CLAVERINGS, Anthony Trollope. Major novel, chronicling aspects of British Victorian society, personalities. 16 plates by M. Edwards; first reprint of full text. 412pp. 5⅜ × 8½. 23464-9 Pa. $6.00

EINSTEIN'S THEORY OF RELATIVITY, Max Born. Finest semi-technical account; much explanation of ideas and math not readily available elsewhere on this level. 376pp. 5⅜ × 8½. 60769-0 Pa. $5.00

COMPUTABILITY AND UNSOLVABILITY, Martin Davis. Classic graduate-level introduction th theory of computability, usually referred to as theory of recurrent functions. New preface and appendix. 288pp. 5⅜ × 8½. 61471-9 Pa. $6.50

THE GODS OF THE EGYPTIANS, E.A. Wallis Budge. Never excelled for richness, fullness: all gods, goddesses, demons, mythical figures of Ancient Egypt; their legends, rites, incarnations, etc. Over 225 illustrations, plus 6 color plates. 988pp. 6⅛ × 9¼. (EBE) 22055-9, 22056-7 Pa., Two-vol. set $20.00

THE I CHING (THE BOOK OF CHANGES), translated by James Legge. Most penetrating divination manual ever prepared. Indispensable to study of early Oriental civilizations, to modern inquiring reader. 448pp. 5⅜ × 8½.
21062-6 Pa. $6.50

THE CRAFTSMAN'S HANDBOOK, Cennino Cennini. 15th-century handbook, school of Giotto, explains applying gold, silver leaf; gesso; fresco painting, grinding pigments, etc. 142pp. 6⅛ × 9¼. 20054-X Pa. $3.50

AN ATLAS OF ANATOMY FOR ARTISTS, Fritz Schider. Finest text, working book. Full text, plus anatomical illustrations; plates by great artists showing anatomy. 593 illustrations. 192pp. 7⅛ × 10¼. 20241-0 Pa. $6.50

EASY-TO-MAKE STAINED GLASS LIGHTCATCHERS, Ed Sibbett, Jr. 67 designs for most enjoyable ornaments: fruits, birds, teddy bears, trumpet, etc. Full size templates. 64pp. 8¼ × 11. 24081-9 Pa. $3.95

TRIAD OPTICAL ILLUSIONS AND HOW TO DESIGN THEM, Harry Turner. Triad explained in 32 pages of text, with 32 pages of Escher-like patterns on coloring stock. 92 figures. 32 plates. 64pp. 8¼ × 11. 23549-1 Pa. $2.95

KEYBOARD WORKS FOR SOLO INSTRUMENTS, G.F. Handel. 35 neglected works from Handel's vast oeuvre, originally jotted down as improvisations. Includes Eight Great Suites, others. New sequence. 174pp. 9⅜ × 12¼.

24338-9 Pa. $7.50

AMERICAN LEAGUE BASEBALL CARD CLASSICS, Bert Randolph Sugar. 82 stars from 1900s to 60s on facsimile cards. Ruth, Cobb, Mantle, Williams, plus advertising, info, no duplications. Perforated, detachable. 16pp. 8¼ × 11.

24286-2 Pa. $2.95

A TREASURY OF CHARTED DESIGNS FOR NEEDLEWORKERS, Georgia Gorham and Jeanne Warth. 141 charted designs: owl, cat with yarn, tulips, piano, spinning wheel, covered bridge, Victorian house and many others. 48pp. 8¼ × 11.

23558-0 Pa. $1.95

DANISH FLORAL CHARTED DESIGNS, Gerda Bengtsson. Exquisite collection of over 40 different florals: anemone, Iceland poppy, wild fruit, pansies, many others. 45 illustrations. 48pp. 8¼ × 11. 23957-8 Pa. $1.75

OLD PHILADELPHIA IN EARLY PHOTOGRAPHS 1839-1914, Robert F. Looney. 215 photographs: panoramas, street scenes, landmarks, President-elect Lincoln's visit, 1876 Centennial Exposition, much more. 230pp. 8⅜ × 11¾.

23345-6 Pa. $9.95

PRELUDE TO MATHEMATICS, W.W. Sawyer. Noted mathematician's lively, stimulating account of non-Euclidean geometry, matrices, determinants, group theory, other topics. Emphasis on novel, striking aspects. 224pp. 5⅜ × 8½.

24401-6 Pa. $4.50

ADVENTURES WITH A MICROSCOPE, Richard Headstrom. 59 adventures with clothing fibers, protozoa, ferns and lichens, roots and leaves, much more. 142 illustrations. 232pp. 5⅜ × 8½. 23471-1 Pa. $3.95

IDENTIFYING ANIMAL TRACKS: MAMMALS, BIRDS, AND OTHER ANIMALS OF THE EASTERN UNITED STATES, Richard Headstrom. For hunters, naturalists, scouts, nature-lovers. Diagrams of tracks, tips on identification. 128pp. 5⅜ × 8. 24442-3 Pa. $3.50

VICTORIAN FASHIONS AND COSTUMES FROM HARPER'S BAZAR, 1867-1898, edited by Stella Blum. Day costumes, evening wear, sports clothes, shoes, hats, other accessories in over 1,000 detailed engravings. 320pp. 9⅜ × 12¼.

22990-4 Pa. $10.95

EVERYDAY FASHIONS OF THE TWENTIES AS PICTURED IN SEARS AND OTHER CATALOGS, edited by Stella Blum. Actual dress of the Roaring Twenties, with text by Stella Blum. Over 750 illustrations, captions. 156pp. 9 × 12.

24134-3 Pa. $8.50

HALL OF FAME BASEBALL CARDS, edited by Bert Randolph Sugar. Cy Young, Ted Williams, Lou Gehrig, and many other Hall of Fame greats on 92 full-color, detachable reprints of early baseball cards. No duplication of cards with *Classic Baseball Cards*. 16pp. 8¼ × 11. 23624-2 Pa. $3.50

THE ART OF HAND LETTERING, Helm Wotzkow. Course in hand lettering, Roman, Gothic, Italic, Block, Script. Tools, proportions, optical aspects, individual variation. Very quality conscious. Hundreds of specimens. 320pp. 5⅜ × 8½.

21797-3 Pa. $4.95

THE RIME OF THE ANCIENT MARINER, Gustave Doré, S.T. Coleridge. Doré's finest work, 34 plates capture moods, subtleties of poem. Full text. 77pp. 9¼ × 12.   22305-1 Pa. $4.95

SONGS OF INNOCENCE, William Blake. The first and most popular of Blake's famous "Illuminated Books," in a facsimile edition reproducing all 31 brightly colored plates. Additional printed text of each poem. 64pp. 5¼ × 7.
22764-2 Pa. $3.50

AN INTRODUCTION TO INFORMATION THEORY, J.R. Pierce. Second (1980) edition of most impressive non-technical account available. Encoding, entropy, noisy channel, related areas, etc. 320pp. 5⅜ × 8½.   24061-4 Pa. $4.95

THE DIVINE PROPORTION: A STUDY IN MATHEMATICAL BEAUTY, H.E. Huntley. "Divine proportion" or "golden ratio" in poetry, Pascal's triangle, philosophy, psychology, music, mathematical figures, etc. Excellent bridge between science and art. 58 figures. 185pp. 5⅜ × 8½.   22254-3 Pa. $3.95

THE DOVER NEW YORK WALKING GUIDE: From the Battery to Wall Street, Mary J. Shapiro. Superb inexpensive guide to historic buildings and locales in lower Manhattan: Trinity Church, Bowling Green, more. Complete Text; maps. 36 illustrations. 48pp. 3⅞ × 9¼.   24225-0 Pa. $2.50

NEW YORK THEN AND NOW, Edward B. Watson, Edmund V. Gillon, Jr. 83 important Manhattan sites: on facing pages early photographs (1875-1925) and 1976 photos by Gillon. 172 illustrations. 171pp. 9¼ × 10.   23361-8 Pa. $7.95

HISTORIC COSTUME IN PICTURES, Braun & Schneider. Over 1450 costumed figures from dawn of civilization to end of 19th century. English captions. 125 plates. 256pp. 8⅜ × 11¼.   23150-X Pa. $7.50

VICTORIAN AND EDWARDIAN FASHION: A Photographic Survey, Alison Gernsheim. First fashion history completely illustrated by contemporary photographs. Full text plus 235 photos, 1840-1914, in which many celebrities appear. 240pp. 6½ × 9¼.   24205-6 Pa. $6.00

CHARTED CHRISTMAS DESIGNS FOR COUNTED CROSS-STITCH AND OTHER NEEDLECRAFTS, Lindberg Press. Charted designs for 45 beautiful needlecraft projects with many yuletide and wintertime motifs. 48pp. 8¼ × 11.
24356-7 Pa. $2.50

101 FOLK DESIGNS FOR COUNTED CROSS-STITCH AND OTHER NEEDLE-CRAFTS, Carter Houck. 101 authentic charted folk designs in a wide array of lovely representations with many suggestions for effective use. 48pp. 8¼ × 11.
24369-9 Pa. $2.25

FIVE ACRES AND INDEPENDENCE, Maurice G. Kains. Great back-to-the-land classic explains basics of self-sufficient farming. The one book to get. 95 illustrations. 397pp. 5⅜ × 8½.   20974-1 Pa. $4.95

A MODERN HERBAL, Margaret Grieve. Much the fullest, most exact, most useful compilation of herbal material. Gigantic alphabetical encyclopedia, from aconite to zedoary, gives botanical information, medical properties, folklore, economic uses, and much else. Indispensable to serious reader. 161 illustrations. 888pp. 6½ × 9¼. (Available in U.S. only)   22798-7, 22799-5 Pa., Two-vol. set $16.45

REASON IN ART, George Santayana. Renowned philosopher's provocative, seminal treatment of basis of art in instinct and experience. Volume Four of *The Life of Reason*. 230pp. 5⅜ × 8. 24358-3 Pa. $4.50

LANGUAGE, TRUTH AND LOGIC, Alfred J. Ayer. Famous, clear introduction to Vienna, Cambridge schools of Logical Positivism. Role of philosophy, elimination of metaphysics, nature of analysis, etc. 160pp. 5⅜ × 8½. (USCO) 20010-8 Pa. $2.75

BASIC ELECTRONICS, U.S. Bureau of Naval Personnel. Electron tubes, circuits, antennas, AM, FM, and CW transmission and receiving, etc. 560 illustrations. 567pp. 6½ × 9¼. 21076-6 Pa. $8.95

THE ART DECO STYLE, edited by Theodore Menten. Furniture, jewelry, metalwork, ceramics, fabrics, lighting fixtures, interior decors, exteriors, graphics from pure French sources. Over 400 photographs. 183pp. 8⅜ × 11¼. 22824-X Pa. $6.95

THE FOUR BOOKS OF ARCHITECTURE, Andrea Palladio. 16th-century classic covers classical architectural remains, Renaissance revivals, classical orders, etc. 1738 Ware English edition. 216 plates. 110pp. of text. 9½ × 12¾. 21308-0 Pa. $11.50

THE WIT AND HUMOR OF OSCAR WILDE, edited by Alvin Redman. More than 1000 ripostes, paradoxes, wisecracks: Work is the curse of the drinking classes, I can resist everything except temptations, etc. 258pp. 5⅜ × 8½. (USCO) 20602-5 Pa. $3.95

THE DEVIL'S DICTIONARY, Ambrose Bierce. Barbed, bitter, brilliant witticisms in the form of a dictionary. Best, most ferocious satire America has produced. 145pp. 5⅜ × 8½. 20487-1 Pa. $2.50

ERTÉ'S FASHION DESIGNS, Erté. 210 black-and-white inventions from *Harper's Bazar*, 1918-32, plus 8pp. full-color covers. Captions. 88pp. 9 × 12. 24203-X Pa. $6.50

ERTÉ GRAPHICS, Erté. Collection of striking color graphics: *Seasons, Alphabet, Numerals, Aces* and *Precious Stones*. 50 plates, including 4 on covers. 48pp. 9⅜ × 12¼. 23580-7 Pa. $6.95

PAPER FOLDING FOR BEGINNERS, William D. Murray and Francis J. Rigney. Clearest book for making origami sail boats, roosters, frogs that move legs, etc. 40 projects. More than 275 illustrations. 94pp. 5⅜ × 8½. 20713-7 Pa. $2.25

ORIGAMI FOR THE ENTHUSIAST, John Montroll. Fish, ostrich, peacock, squirrel, rhinoceros, Pegasus, 19 other intricate subjects. Instructions. Diagrams. 128pp. 9 × 12. 23799-0 Pa. $4.95

CROCHETING NOVELTY POT HOLDERS, edited by Linda Macho. 64 useful, whimsical pot holders feature kitchen themes, animals, flowers, other novelties. Surprisingly easy to crochet. Complete instructions. 48pp. 8¼ × 11. 24296-X Pa. $1.95

CROCHETING DOILIES, edited by Rita Weiss. Irish Crochet, Jewel, Star Wheel, Vanity Fair and more. Also luncheon and console sets, runners and centerpieces. 51 illustrations. 48pp. 8¼ × 11. 23424-X Pa. $2.50

YUCATAN BEFORE AND AFTER THE CONQUEST, Diego de Landa. Only significant account of Yucatan written in the early post-Conquest era. Translated by William Gates. Over 120 illustrations. 162pp. 5⅜ × 8½. 23622-6 Pa. $3.50

ORNATE PICTORIAL CALLIGRAPHY, E.A. Lupfer. Complete instructions, over 150 examples help you create magnificent "flourishes" from which beautiful animals and objects gracefully emerge. 8⅛ × 11. 21957-7 Pa. $2.95

DOLLY DINGLE PAPER DOLLS, Grace Drayton. Cute chubby children by same artist who did Campbell Kids. Rare plates from 1910s. 30 paper dolls and over 100 outfits reproduced in full color. 32pp. 9¼ × 12¼. 23711-7 Pa. $3.50

CURIOUS GEORGE PAPER DOLLS IN FULL COLOR, H. A. Rey, Kathy Allert. Naughty little monkey-hero of children's books in two doll figures, plus 48 full-color costumes: pirate, Indian chief, fireman, more. 32pp. 9¼ × 12¼.
24386-9 Pa. $3.50

GERMAN: HOW TO SPEAK AND WRITE IT, Joseph Rosenberg. Like *French, How to Speak and Write It.* Very rich modern course, with a wealth of pictorial material. 330 illustrations. 384pp. 5⅜ × 8½. (USUKO) 20271-2 Pa. $4.75

CATS AND KITTENS: 24 Ready-to-Mail Color Photo Postcards, D. Holby. Handsome collection; feline in a variety of adorable poses. Identifications. 12pp. on postcard stock. 8¼ × 11. 24469-5 Pa. $2.95

MARILYN MONROE PAPER DOLLS, Tom Tierney. 31 full-color designs on heavy stock, from *The Asphalt Jungle, Gentlemen Prefer Blondes,* 22 others. 1 doll. 16 plates. 32pp. 9⅜ × 12¼. 23769-9 Pa. $3.50

FUNDAMENTALS OF LAYOUT, F.H. Wills. All phases of layout design discussed and illustrated in 121 illustrations. Indispensable as student's text or handbook for professional. 124pp. 8⅜ × 11. 21279-3 Pa. $4.50

FANTASTIC SUPER STICKERS, Ed Sibbett, Jr. 75 colorful pressure-sensitive stickers. Peel off and place for a touch of pizzazz: clowns, penguins, teddy bears, etc. Full color. 16pp. 8¼ × 11. 24471-7 Pa. $2.95

LABELS FOR ALL OCCASIONS, Ed Sibbett, Jr. 6 labels each of 16 different designs—baroque, art nouveau, art deco, Pennsylvania Dutch, etc.—in full color. 24pp. 8¼ × 11. 23688-9 Pa. $2.95

HOW TO CALCULATE QUICKLY: RAPID METHODS IN BASIC MATHE-MATICS, Henry Sticker. Addition, subtraction, multiplication, division, checks, etc. More than 8000 problems, solutions. 185pp. 5 × 7¼. 20295-X Pa. $2.95

THE CAT COLORING BOOK, Karen Baldauski. Handsome, realistic renderings of 40 splendid felines, from American shorthair to exotic types. 44 plates. Captions. 48pp. 8¼ × 11. 24011-8 Pa. $2.25

THE TALE OF PETER RABBIT, Beatrix Potter. The inimitable Peter's terrifying adventure in Mr. McGregor's garden, with all 27 wonderful, full-color Potter illustrations. 55pp. 4¼ × 5½. (Available in U.S. only) 22827-4 Pa. $1.75

BASIC ELECTRICITY, U.S. Bureau of Naval Personnel. Batteries, circuits, conductors, AC and DC, inductance and capacitance, generators, motors, trans-formers, amplifiers, etc. 349 illustrations. 448pp. 6½ × 9¼. 20973-3 Pa. $7.95

THE PRINCIPLE OF RELATIVITY, Albert Einstein et al. Eleven most important original papers on special and general theories. Seven by Einstein, two by Lorentz, one each by Minkowski and Weyl. 216pp. 5⅜ × 8½.   **60081-5 Pa. $4.00**

PINEAPPLE CROCHET DESIGNS, edited by Rita Weiss. The most popular crochet design. Choose from doilies, luncheon sets, bedspreads, apron—34 in all. 32 photographs. 48pp. 8¼ × 11.   **23939-X Pa. $2.00**

REPEATS AND BORDERS IRON-ON TRANSFER PATTERNS, edited by Rita Weiss. Lovely florals, geometrics, fruits, animals, Art Nouveau, Art Deco and more. 48pp. 8¼ × 11.   **23428-2 Pa. $1.95**

SCIENCE-FICTION AND HORROR MOVIE POSTERS IN FULL COLOR, edited by Alan Adler. Large, full-color posters for 46 films including *King Kong, Godzilla, The Illustrated Man,* and more. A bug-eyed bonanza of scantily clad women, monsters and assorted other creatures. 48pp. 10¼ × 14¼. **23452-5 Pa. $8.95**

TECHNICAL MANUAL AND DICTIONARY OF CLASSICAL BALLET, Gail Grant. Defines, explains, comments on steps, movements, poses and concepts. 15-page pictorial section. Basic book for student, viewer. 127pp. 5⅜ × 8½.
    21843-0 Pa. $2.95

STORYBOOK MAZES, Dave Phillips. 23 stories and mazes on two-page spreads: *Wizard of Oz, Treasure Island, Robin Hood,* etc. Solutions. 64pp. 8¼ × 11.
    23628-5 Pa. $2.25

PUNCH-OUT PUZZLE KIT, K. Fulves. Engaging, self-contained space age entertainments. Ready-to-use pieces, diagrams, detailed solutions. Challenge a robot, split the atom, more. 40pp. 8¼ × 11.   **24307-9 Pa. $3.50**

THE HUMAN FIGURE IN MOTION, Eadweard Muybridge. Over 4500 19th-century photos showing stopped-action sequences of undraped men, women, children jumping, running, sitting, other actions. Monumental collection. 390pp. 7⅞ × 10⅝.   **20204-6 Clothbd. $18.95**

PHOTOGRAPHIC SKETCHBOOK OF THE CIVIL WAR, Alexander Gardner. Reproduction of 1866 volume with 100 on-the-field photographs: Manassas, Lincoln on battlefield, slave pens, etc. 224pp. 10⅝ × 8¼.   **22731-6 Pa. $7.95**

FLORAL IRON-ON TRANSFER PATTERNS, edited by Rita Weiss. 55 floral designs, large and small, realistic, stylized; poppies, iris, roses, etc. Victorian, modern. Instructions. 48pp. 8¼ × 11.   **23248-4 Pa. $1.95**

AUTOBIOGRAPHY: The Story of My Experiments with Truth, Mohandas K. Gandhi. Boyhood, legal studies, purification, the growth of the Satyagraha (nonviolent protest) movement. Critical, inspiring work of the man who freed India. 480pp. 5⅜ × 8½.   **24593-4 Pa. $6.95**

ON THE IMPROVEMENT OF THE UNDERSTANDING, Benedict Spinoza. Also contains *Ethics, Correspondence,* all in excellent R Elwes translation. Basic works on entry to philosophy, pantheism, exchange of ideas with great contemporaries. 420pp. 5⅜ × 8½.   **20250-X Pa. $5.95**

Prices subject to change without notice.

Available at your book dealer or write for free catalog to Dept. GI, Dover Publications, Inc., 31 East 2nd St. Mineola, N.Y. 11501. Dover publishes more than 175 books each year on science, elementary and advanced mathematics, biology, music, art, literary history, social sciences and other areas.